T0137378

NEW APPROACHES TO FOOD-SAFETY ECONOMICS

Wageningen UR Frontis Series

VOLUME 1

Series editor:

R.J. Bogers
Frontis – Wageningen International Nucleus for Strategic Expertise, Wageningen University and Research Centre, The Netherlands

NEW APPROACHES TO FOOD-SAFETY ECONOMICS

Edited by

A.G.J. VELTHUIS

Farm Management Group,
Wageningen University, The Netherlands

L.J. UNNEVEHR

Department of Agricultural & Consumer Economics,
University of Illinois at Urbana-Champaign, U.S.A.

H. HOGEVEEN

Farm Management Group,
Wageningen University, The Netherlands

and

R.B.M. HUIRNE

Farm Management Group,
Wageningen University, The Netherlands

KLUWER ACADEMIC PUBLISHERS

DORDRECHT / BOSTON / LONDON

A C.I.P. Catalogue record for this book is available from the Library of Congress.

ISBN 1-4020-1425-2 (HB)
ISBN 1-4020-1426-0 (PB)

Published by Kluwer Academic Publishers,
P.O. Box 17, 3300 AA Dordrecht, The Netherlands.

Sold and distributed in North, Central and South America
by Kluwer Academic Publishers,
101 Philip Drive, Norwell, MA 02061, U.S.A.

In all other countries, sold and distributed
by Kluwer Academic Publishers,
P.O. Box 322, 3300 AH Dordrecht, The Netherlands.

Printed on acid-free paper

Printed in the Netherlands.

Contents

Transparency in intra-EU and international trade

Preface

The emergence of new pathogens, changes in the food system, and increased food-product trade all led to increased attention to food-safety issues during the 1990s. A number of countries implemented new food-safety regulations, including the EU Food Safety Law of 2002. Food industry is increasing efforts to certify food safety and reassure consumers. In this context, economic research during the last decade has begun to explore the costs, benefits, and trade-offs in food-safety improvement.

Food-safety economics is a new research field, which needs a solid framework of concepts, procedures and data to support the decision-making process in food-safety improvement. Food safety is a theme that plays at many levels in the community: at the consumer level, at the farm or business level, at the supply-chain level, nationally and internationally. Moreover, there are a lot of aspects involved in food safety and still a lot of questions are unanswered.

In order to frame a research agenda to meet the challenges facing industry and government, invited experts from around the world were invited to a workshop in April 2002 at Wageningen University and Research Centre in order to review the state of the art. The purpose of the workshop was to learn where and how economics can contribute to emerging policies and strategies for food-safety improvement. In the EU, the public desire for greater food-safety assurance is evoking response from both the food industry and policymakers. These responses take place within the context of changes in EU membership, in agricultural policies, and in scientific knowledge about risks. The primary focus of the workshop was on food-safety economics for animal products.

Format of the Book

The workshop was organized around the key principles outlined in the EU White Paper on Food Safety, as these will guide future developments in the EU and other parts of the world. These principles include a focus on consumer welfare, responsibility for food safety from farm to table, the use of risk analysis to design standards, prevention of hazards through the use of HACCP, implementation of traceability to ensure monitoring, and transparency of standards in international trade.

The chapters in the book are organized into four sections around these themes. Within each theme area, the chapters include different perspectives and approaches to the issues. The papers address the state of the art, areas for new research, and areas where knowledge from other scientific disciplines is needed. At the workshop, general discussions among all participants and identification of new research directions took place within each theme area. The results of these discussions are reported in the overview chapter, which also provides a summary of the chapters and presentations.

The organizers

The workshop was organized by Frontis – Wageningen International Nucleus for Strategic Expertise, and the Farm Management Group of the Department of Social Sciences, both at Wageningen University and Research Centre, Wageningen, The Netherlands.

A.G. J. Velthuis et al. (eds), New Approaches to Food-Safety Economics, vii.

Acknowledgements

We would like to thank the authors for their contributions, Henny Wilpshaar, Marian Jonker and Wendelien Ordelman for making the organizational arrangements for the workshop, Annet Velthuis and Henk Hogeveen for editing the manuscripts, Hugo Besemer and Pauline van Vredendaal for facilitating the lay-out process and Wageningen University and Research Centre for financial support.

Wageningen/Illinois, December 2002

Ruud B.M. Huirne
Laurian J. Unnevehr
Robert J. Bogers

1

New approaches to food-safety economics: overview and new research directions

Laurian J. Unnevehr[*] *and Ruud B.M. Huirne*[#]

Introduction

The papers in this volume were first presented at a Frontis workshop on New Approaches to Food-Safety Economics, held at Wageningen, the Netherlands, 14-17 April 2002. The purpose of the symposium was to learn where and how economics can contribute to emerging policies and strategies for food-safety improvement. In the EU, public desire for greater food-safety assurance is evoking response from both the food industry and policymakers. These responses take place within the context of changes in EU membership, in agricultural policies, and in scientific knowledge about risks. In order to frame a research agenda to meet the challenges facing industry and government, invited experts from around the world convened to review the state of the art. The primary focus of the workshop was on food-safety economics for animal products.

The papers in this volume are written on invitation and are organized around the key principles outlined in the EU White Paper on Food Safety, as these will guide future developments in the EU. These principles include a focus on consumer welfare, responsibility for food safety from farm to table, the use of risk analysis to design standards, prevention of hazards through the use of HACCP, implementation of traceability to ensure monitoring, and transparency of standards in international trade.

In the following overview, we provide the highlights and important insights from the presentations and discussions. We conclude with the major research areas identified for future work.

Consumer Health and Welfare

Consumer health and welfare is the ultimate goal of food-safety improvement. There are several approaches to measuring consumer welfare, and to understanding consumer perceptions and behavior. The first group of papers provides an overview of these approaches, drawing on methods from economics, marketing, and public health. How to measure the value of food safety improvements, understand consumer perceptions and market behavior, and how to set public-health goals were among the issues addressed.

In Shogren's paper, the economics literature on benefits estimation is reviewed, with particular attention to WTP estimation. He provides several insights from experimental auctions in the US Consumers tend to underestimate risk; they value

[*] *Department of Agricultural & Consumer Economics, University of Illinois at Urbana-Champaign, USA*

[#] *Farm Management Group, Wageningen University, Wageningen, The Netherlands*

A.G. J. Velthuis et al. (eds), New Approaches to Food-Safety Economics, 1-8.

food safety generally but not specific hazards; and negative news has more impact than positive news on consumer valuation. In general, people are willing to pay for higher safety than the market currently provides. The findings lead to some lessons for technical scientists. Economics should be part of risk assessment because economic decisions partly determine risk. The public has different risk perceptions from experts: they underestimate both low-probability and high-probability risks, but overestimate the mid-range of risk probability.

Verbeke reviews several studies that examined how Belgian consumers perceive food safety, how they react to new information, and how they alter food consumption in response to new information. He finds that the established trend towards reduced beef consumption was strengthened by reaction to the BSE crisis and that negative perceptions of poultry resulted from the dioxin crisis. The degree to which consumers watch TV had a major impact on behavior and perceptions. His estimates show that five items of good news are equal in impact to one item of bad news. He proposes that consumer trust may be restored through labeling and traceability.

Henken and his co-authors report estimates of the number of food-borne-illness cases in the Netherlands. These data show that Campylobacter and viruses more important than Salmonella, and warrant more attention. Risk assessment has been carried out for steak tartare and for Campylobacter in the Netherlands. One important issue in such assessments is establishing the dose–response function for microbial pathogens, and he reports substantial progress in their identification. From a public-health perspective, the Disability-Adjusted Life Year (DALY), measures the impact of food-borne illness. The DALYs saved from interventions to reduce risk can be compared with cost to find the most cost-effective interventions.

The conference discussion focused on the issues of communication and perception. The need to understand what information people would like to have and whether they trust certain sources of information are important areas for research. The role of so-called experts was raised, and whether they should be the source of information was questioned. There is no one scientific truth, and furthermore technical experts are not trained in risk communication. The finding that consumers tend to place more weight on bad news than on good news was seen as a challenge for industry. Finally, the issue of whether we should really invest more resources in food safety and how to measure the benefits of further investment was raised.

Traceability and Certification in the Supply Chain

Traceability and certification are processes for managing and marketing food quality, including food safety. While certification has been used by many firms for some time, traceability is a relatively new concept in food safety. The costs and benefits of these two approaches for achieving food safety are not well understood. The papers in this section provide an overview of the current systems in use, and the economic and managerial questions to be answered about such systems.

Meuwissen et al. provide an overview of traceability and certification systems. Regulatory and private standards provide the goals for private firms, which may use traceability systems to help meet those goals. Certification ensures that management systems work as they are supposed to. There are three possible goals for a traceability system, including establishment of consumer confidence, avoidance of liability, and improvement in recall efficiency. There are different ways of organizing traceability systems in terms of how information is shared along the chain, and the choice of system may be influenced by the relative importance of the different goals.

Certification may be required by a customer, may result in better prices, or may be required by financial institutions. It is usually carried out by a third party, and may or may not certify to an accredited standard. Both traceability and certification may have implications for the organization of the supply chain as they are easier to accomplish when there is one chain director or an integrated system.

Schiefer's paper approaches the issues of traceability and certification from a management perspective. Research in this field has shown that quality assurance should be a dynamic forward-looking process seeking to improve quality. But the current proposals for traceability and certification in the agricultural sector are primarily defensive, and not forward-looking. Closed supply chains would provide the basis for higher quality and incentives for continual improvement. Network systems that include all producers build on generally acceptable but low levels of quality and do not provide incentives for improvement. Thus, current initiatives in Germany may not provide for better quality and safety and may ultimately lead to reduced consumer confidence.

Den Hartog presented the agribusiness perspective of a firm involved in fish, poultry, and pig production chains.[i] Agribusiness faces issues arising from recent food-safety incidents, media coverage, and new regulation. While a defensive reaction may be natural, his firm, Nutreco, is approaching food safety in a pro-active manner. Part of establishing reliability is to provide for traceability in their integrated production chains. The poultry production chain from feed to meat is the first place that this system will be implemented. A database will be created to provide information about all of the ingredients and processes used in production. Ultimately this should lead to greater ability to manage safety in an efficient manner.

The discussion opened with concern about how to set the performance standards for traceability and certification systems. Such systems will be meaningless if consumers do not understand what is achieved in terms of food safety. A second important concern was the implication of traceability for small farms, who may be excluded because they have high costs of supplying information. The discussion also explored the issue of liability, and distinguished between certification, which can reduce liability, and traceability, which may make it easier to identify the liable party. The economic value of certification or traceability may arise from being able to reduce costs later in the chain. The ease of implementing such systems may be enhanced by new information technologies. Finally, well organized integrated production chains are seen as the future in animal agriculture.

Farm-to-Table Risk Analysis

The nature of food-borne hazards makes it desirable to analyse risks in a farm-to-table framework. Many food-safety hazards can enter the food production chain at mulitple points and can multiply or cross-contaminate other products once present. Thus a farm-to-table approach allows identification of the most effective points for intervention. Integrating economics into this framework is challenging, but can allow identification of the most cost-effective interventions. The papers in this session provide different perspectives on farm-to-table analysis of risks and costs, and how this might support decision making in the public and private sectors.

Jensen defines the costs of food safety as real resource costs used in hazard prevention, social-welfare losses arising from changes in market prices, and transitional costs arising from firm closings or reorganization. The literature on costs of food-safety improvement, primarily US studies of HACCP for meat and poultry, shows that there are rising marginal costs associated with higher levels of safety. Because several interventions are often possible, there is a frontier of dominant cost-effective interventions. Analysis of interventions in a farm-to-table framework allows identification of the most cost-effective points of intervention in the supply chain and how later interventions may support earlier ones. Economic incentives to provide food safety are influenced by system connectivity, which spreads breakdowns in safety through the supply network in particular paths. Consideration of these system connections may lead to design of better incentives for food safety.

Lund's paper reports on a holistic effort to analyse the economics of food safety throughout the entire production chain. This is being undertaken for three animal products in Denmark. Through scenario modeling, the implications of alternative marketing strategies and food policies will be understood. Lund identifies several challenges in this analysis, including the difficulty of valuing the benefits of food safety, modeling connections in the food chain, and fully identifying the underlying risk model. In terms of policy, ensuring that feedback mechanisms along the food chain lead to improvement will be a challenge in implementation HACCP from farm to table.

Stark's presentation reviewed the evolution and use of risk analysis.[ii] It has been used in import–export risk management for veterinary applications, and is increasingly applied to food-safety issues. A model developed in Denmark has been used to assess the risk of Salmonella contamination along the pork supply chain and as a decision-support tool for Danish pig producers. Integration of the Danish pork production chain makes it feasible for actions to be undertaken with this systems approach. Stark identified a number of issues in data and methods for building risk-assessment models. A particular difficulty is that such models often must rely on expert opinion. There is a need to elicit such opinions carefully and to follow up through validation of the model with actual surveillance data.

The discussion focused on the goals and limitations of risk assessment, and the role of economics. Defining food-safety goals is seen as a challenge for policymakers. Presumably it may be defined as a specific reduction in prevalence, but it may also be defined as a dynamic goal of continuous improvement. Economics can play a role in helping to define goals, by equating marginal costs with marginal benefits. The integration of economics into risk-assessment models may provide decision support for producers to choose between different interventions to meet specific standards. The scope of risk assessment was also discussed. If a risk assessment is truly oriented towards public health, then current models must be extended to include consumer behavior and the actual endpoint of the food chain. This might help in evaluation of consumer-education efforts. Finally, there was discussion of whether a systems approach is practical for either modeling or implementation of control options. Systems approaches require extensive data and increasingly complex models. Implementation of systems approaches for food-safety control may not be possible without an integrated food supply chain.

Transparency in International Trade

International trade in food products is growing rapidly, allowing consumers in many countries greater variety of foods at lower prices. At the same time, such trade can introduce new or different sources of risks. Food-safety standards are increasingly strict in many high-income countries, and can be a non-tariff barrier to trade. The Sanitary and Phytosanitary (SPS) Agreement of the World Trade Organization

(WTO) provides a framework of principles for such standards. While allowing countries to set their own standards, it ensures that such standards do not unduly restrict trade. A key principle is transparency in standards. As noted by Magelhães, this allows government authorities and private firms to identify and deal with potential market-access problems. The papers in this session explore these complex trade issues from different perspectives, including the implications of trade agreements for regulation and the performance of the SPS agreement in ensuring market access.

Marette et al. focus on two forces shaping evolving food-safety regulation. First, consumer trust has been shaken in Europe by food-safety crises. Second, international trade agreements shape new regulations through requirements for science-based risk assessment and transparent standards, as well as the recognition of international minimum standards set through the Codex Alimentarius. In the new EU food-safety law, there is increased emphasis on command and control as well as information-based approaches, such as labeling. More incentive-based approaches, such as product-liability laws, have not been emphasized. Differences between the emerging EU and US approaches are evident in the regulation of GMOs and the establishment of liability. Such differences may lead to further tensions in international markets. Marette et al. propose that the EU should give further consideration to the use of product-liability laws, which are emerging for environmental issues and may be reinforced through the adoption of traceability systems.

Wilson's chapter starts from the widespread observation that domestic regulation may pose non-tariff barriers to trade, and that food safety-regulations have impeded trade. However, as he notes, there are few empirical studies actually measuring the impact of new regulations on traded quantities. He then reviews in depth some selected research literature that measures the impact of food-safety regulations on quantities traded. In general, these studies find that more stringent food-safety regulations tend to reduce traded quantities. The examples include several changes in EU regulations that set more stringent standards than those recognized in the Codex Alimentarius. The impacts tend to be severe for less developed countries, where it may be more difficult to meet higher standards. Thus, there is tension between balancing national interests in higher standards with reducing barriers to trade.

Magelhães' chapter provides background on the meaning of transparency under the SPS agreement and data regarding how well countries have complied with the agreement. An important part of the agreement is the notification system, under which countries are required to post proposed changes in SPS measures. This provides other WTO members with knowledge of pending changes and allows them the opportunity to raise concerns. Statistics on the use of notification systems show that transparency provisions are increasing dialogue and communication among member nations. A specific example explored in the presentation was the reaction of exporting countries to proposed EU revisions in aflatoxin standards.

The discussion began by considering whether or not the SPS agreement leads to improved trade, welfare, and development. Transaction costs of meeting SPS obligations, such as participation in the Codex Alimentarius, are very high for less developed countries. But on the other hand, this agreement has clearly increased transparency among trading partners. More trade ultimately leads to greater wealth, with corresponding benefits for food safety. Another important issue is that meeting higher food-safety standards may be difficult for small farms or small firms. Expanded trade and higher standards may lead to a more concentrated market structure or to increased foreign direct investment in exporting countries. The role of the private sector, especially European retailers, is very important in setting the private food-safety standards for imports. These private sector actions are beyond the jurisdiction of the SPS agreement, but may determine international trade in the future.

As a result of this discussion, the conference closed with invited presentations from two participants representing different international trade perspectives. De Haan's chapter reviews the efforts to the World Bank to improve food safety in developing countries. These include policy dialogue, institution-building support, and infrastructure support. Such activities are seen as essential to the development of high-valued export industries in developing economies. Nagamatsu and Matsuki provided some perspective from Japan, where concerns about food safety mirror those in other industrialized countries. As in Europe and the US, there are new food-safety laws and expanded food-safety regulation. Japan is seeing the emergence of integrated, certified food supply chains. The development of these supply chains takes place within the traditional framework of producer cooperatives that dominate markets in Japan. These two chapters provide additional context for the other chapters in the book, which primarily cover research carried out in the US and Europe.

New Research Directions Identified

Future directions for research were identified in all four areas. A unifying theme is the need for cooperation between technical and social scientists to answer important questions. There is some degree of overlap among the four research topics, reflecting the integrated nature of risk along the food supply chain. Thus, continued exchange of ideas will provide better solutions to food-safety problems. This includes exchanges among social and technical scientists, as well as among researchers working at different aspects of the food system.

Risk communication

Conference participants were particularly concerned about the inability of technical scientists and economists to understand some aspects of consumer behavior. There is a need for integration of emerging fields in psychology and consumer marketing with economic analyses of market behavior and WTP. Such research might address how and why consumers choose food products and food supply sources; how and why they undertake risk-mitigation measures in preparing or choosing food; and how best to tailor risk-reduction activities to reduce the incidence of food-borne illness. The latter questions relate the need identified for extending risk-assessment models that they truly encompass the system from farm all the way to table.

Developing guidelines for traceability systems

In order to facilitate the development of traceability systems, research is needed to develop the guidelines for such systems, including the design of incentives and establishment of liability. Cost–benefit analysis of traceability systems would be part of this research, to see whether the costs are justified. As part of such analysis, benefits would be broadly defined to include reductions in transaction costs, reductions in administrative costs of regulation, and gains in system efficiency. The need for differentiated systems, such as a government-mandated minimum system and stricter private systems, would be explored. Such research might compare the experiences in countries such as the United Kingdom, where the Food Safety Act of 1990 spurred supply-chain coordination, and Germany, where coordination is less well developed. Such research might also draw on the expertise in economics of animal health in the Netherlands, where traceability has long-established use in animal-disease control.

Integrating economics into farm-to-table risk assessment

Research is needed to understand the costs and benefits of different levels of food safety or among alternative standards. In order to accomplish this, an economic model will be integrated into an existing risk model. This would also allow identification of the distribution of intervention costs along the production chain. Exploring the integration of economics into risk models will also provide insight into how economic factors such as farm size influence risk and whether it is necessary to consider them in risk assessment. An extension of this work would compare the costs and benefits of food safety in alternative production and supply chains. This might compare conventional production with organic production or with an integrated chain. Insights into trade issues might be gained from comparing costs and benefits in the chain from a third-country supplier to EU consumers with the domestic production chain.

Encouraging pro-active risk management in international trade

As international trade grows, the nature and incidence of risks in the food system become a shared concern among countries. Thus, risk reduction may be an international public good. There may be value in coordinated activities to reduce risks, which could promote both trade and food safety. Economists can provide analysis of where such coordination has value and how to address issues arising from the distribution of costs and risks among countries. This is closely tied to the concerns raised in discussion about the value of international standards and the impact of new regulations on developing countries. For example, are harmonized standards the best way to facilitate both trade and risk reduction? Should importing countries compensate exporting countries for investments that reduce food hazards? These types of questions have not been explored in the international trade literature and would facilitate the market adjustments and trade negotiations currently underway in response to higher food-safety standards.

Conclusions

The papers in this volume show the depth and breadth of this emerging field in agricultural economics. Three themes are demonstrated. First, a wide variety of existing economic methods can be applied to different questions in food-safety economics. These include experimental auctions, market models, simulation models, and welfare analysis. Thus, the scope of food-safety issues demands the application of many different kinds of economics methods.

Second, the benefits of multi-disciplinary collaboration are clear. Such collaboration is not new to agricultural and resource economists, but food-safety issues will further challenge economists. Understanding food safety within complex biological and market systems will require new investments in the development of data and appropriate models that integrate technical and social science. Experiences from past efforts in modeling animal health and non-point-source pollution will be relevant, and provide examples for food-safety economics.

Third, these papers demonstrate that parallel trends are underway in the industrialized countries. Consumers are increasingly concerned about food-safety issues, new regulations are emerging to raise standards, and food suppliers are responding with increased quality control and certification. As these trends continue, economists can aid decision makers in understanding the emerging costs, benefits, and trade-offs. International co-operation among researchers will enhance the ability of economists to provide meaningful answers, and in this context, the workshop reported here represents an important step forward.

---

[i] This presentation is not included in this volume.
[ii] This presentation is not included in this volume.

CONSUMER HEALTH AND WELFARE

2

Food-safety economics: consumer health and welfare

Jason F. Shogren[*]

Introduction

Eating offers pleasure at the risk of future pain. This truism holds today more than ever with our increasing ability to detect and identify food-borne illness. Well-publicized outbreaks of cholera, *Salmonella, Listeria monocytogenes*, and *E. Coli 0157:H7* have made people aware that food-borne disease makes a lot of us sick every year. An estimated 76 million illnesses annually in the United States alone, with over 300,000 hospitalizations and 5,000 deaths, imposing an estimated cost in the tens of billions of dollars (Crutchfield et al. 1997; Mead et al. 1999). For example, Buzby and Roberts (Buzby and Roberts 1996) estimated that for six bacterial pathogens, the costs of human illness are estimated to be US$ 9.3 - 12.9 billion annually. Of these costs, US$ 2.9 - 6.7 billion are attributed to food-borne bacteria. One estimate suggests 1 out of 3 consumers in industrialized nations suffers from known and newly recognized food-borne diseases each year (*Food safety - a worldwide public health issue* 2000). And if one looks globally we might also note that: "hundreds of millions of people around the world fall sick as a result of consuming contaminated food and water.... Children under five still suffer an estimated 1.5 billion annual episodes of diarrhea, which result in more than three million premature deaths" (Brundtland 2001).

Many experts anticipate that the risks posed by food-borne disease will increase before they fall. Risks worsen as environmental and demographic conditions change. Sources of new risks include the climate, microbial systems, drinking-water supplies, sanitation, aging, urbanization, migration, consumption habits, tourism, and the mass production and international trade in food and feed (Kaferstein and Abdussalam 1999). The risks are prompting people to demand additional investment, both in the private and public sector, in processes and technologies that will continue to produce inexpensive food but with fewer food-borne risks (e.g., HACCP; irradiation, e.g. (Buzby and Roberts 1996; Lutter 1999; Unnevehr and Jensen 1999; Shogren et al. 1999). Policy experts stress the need to protect the vulnerable – infants and children, pregnant women, the undernourished, the elderly, and the immuno-compromised.

Policymakers in many developed nations have responded to these concerns by publicly committing to strengthen existing programs and to create new policies for safer food. In the United States for example, the Clinton Administration through Executive Order 13100 reinvigorated the question of food safety by introducing the President's Food-safety Initiative (FSI) in 1997, and the President's Council on Food Safety in 1998. The US$ 43-million FSI program focused on reducing the number of illnesses caused by microbial contamination through improved identification and control, enhanced surveillance, and better risk communication and education (Miller and Altekruse 1998). The Council recently unveiled its 2001 Food-safety Strategic

[*] *Department of Economics and Finance, University of Wyoming, Laramie, WY 82071-3985, USA*

A.G. J. Velthuis et al. (eds), New Approaches to Food-Safety Economics, 11-20.

Plan, which promotes science-based systems, prevention, public participation, and setting priorities based on comparative risk analysis. Objective #5 of their plan explicitly calls attention to prioritizing efforts based on the risk-benefit trade-offs: "with limited resources and time, the scientific community must prioritize its efforts to realize its fullest potential. The most significant food-safety problems must be identified and addressed in a manner that enhances public health. Research must be focused and coordinated to avoid duplicative efforts and maximize its benefits" (*Food safety strategic plan* 2001).

Constrained budgets and increased fiscal accountability prevent a policymaker from reducing all food-borne risk to all individuals. Deciding which risks to reduce and by how much requires evaluation of each new or revised regulation. Comparability of value across all sectors of the economy requires that policymakers rank regulatory alternatives in terms of a common unit. Arguably, the most common denominator is money, or monetary equivalence. Risk valuation systematically evaluates each regulation by estimating the monetary value – both benefits and costs – of a reduction in risk from unsafe food. Herein we briefly explore issues in how rational people might value a reduction in risk from food-borne pathogens and other food technologies, and economic methods to measure this value.

Valuing the costs and benefits of risk

Valuing the costs and benefits of reduced risk is formidable and controversial. While measuring the costs to control risk is relatively straightforward, the benefits are a challenge to quantify. Problems arise because goods associated with reduced risk – death and injury – remain unpriced by collective agency action. Stores and restaurants often do not like to market "safer food" because to do so would suggest that their food might otherwise be "unsafe".

Valuing risk reductions requires that we value death and illness. These efforts give rise to the loaded term: "the value of life". The idea of a monetary value of life, or more correctly the value of reduced mortality risk, raises more than a few eyebrows (Schelling 1984; Viscusi 1992). Ethical and moral beliefs often force a person to balk at the idea. But our everyday choices put a value on life, whether we explicitly quantify it or not. Whenever a policy change is enacted or whenever the status quo remains, life and limb are implicitly valued. For example, a North Carolina hospital once refused to spend US$ 150 per healthcare worker for an inoculation against hepatitis B. Given the workers odds of catching the disease, the hospital had implicitly placed a relatively low value on life. Making explicit what we do implicitly provides information about the economic value of reduced statistical risk.

How do we value the welfare gains from a reduction in risk? Holding the level of self-protection constant, the traditional answer is that the value of risk reduction equals:

$$Value\ of\ risk\ reduction = \frac{Willingness\ to\ pay\ for\ risk\ reduction}{Exogenous\ Change\ in\ risk}.$$

Rational risk policy says that a person's value for a risk reduction equals his or her maximum willingness to pay to increase the chances to stay healthy, conditional of his or her previous private actions to reduce risk. For example, suppose a person was willing to pay US$ 6 to reduce the risk of death to 1 life in 1,000,000 from 4 lives in 1,000,000 – a 3 in 1 million-risk reduction. The value of life is then

$$\text{US\$}\,2{,}000{,}000 = \frac{\text{US\$}\,6 \cdot 1{,}000{,}000}{3}.$$

If the person is willing to pay US$ 0.60, the implied value of life would be US$ 200,000; if the person paid US$ 60, the value is US$ 20 million. This ex ante willingness to pay has been called the *option price*. The option price is the maximum a person is willing to pay that keeps him indifferent between the gamble and the next best alternative.

What methods exist to measure actually the value of risk reduction? The literature on rational risk valuation has developed two general approaches to measuring the economic benefits of reduced risk: the human-capital and willingness-to-pay approaches. *The human-capital approach* values risk reductions by examining a person's lifetime earnings and activities. The value of a risk reduction is the gain in future earning and consumption. The value of saving a life is often calculated as what the individual contributes to society through the net present value of future earnings and consumption. The human-capital approach has an advantage in that it is actuarial, i.e., it uses full age-specific accounting to evaluate risk reductions. A major drawback of the approach is that it assigns lower values to the lives of women and minorities, and zero value to retired individuals. The approach also lacks justification based on traditional economic welfare theory. For this reason, economists have downplayed the human-capital method in favor of the willingness-to-pay approach (Buzby et al. 1999)).

Economists have advocated the *willingness-to-pay approach* since it is based on the theory of welfare economics. Welfare economics lays the foundation for estimating the value of risk reduction. People value risk reduction if it leads to a greater level of utility or welfare. The welfare change is measured by the maximum that the average person would be willing to pay to reduce risk or the minimum compensation he or she would be willing to accept for an increase in risk. Economists then use this willingness to pay or accept to estimate the implied value of life and limb. Although far from perfect, economists argue that the willingness-to-pay approach is preferable to the alternative – many believe it is better to have a rough estimate of a well-grounded theory than a precise estimate of a questionable one (Kuchler and Golan 1999). One can reveal this value indirectly by teasing out the implied willingness-to-pay values from real choices within market settings or one can directly estimate values by asking people what they would be willing to pay for a change in risk. See Freeman (Freeman 1993) for a good general overview on non-market valuation and see Caswell (Caswell 1995) for specific case studies using standard valuation methods for food-safety work.

One method we have developed over the last decade to value the willingness to pay for reductions in food-borne illness risk is experimental auction. Over a decade ago, Dermot Hayes, Sean Fox and I became interested in how consumers would react to food safety and new food technologies (Shogren et al. 1994; Shogren et al. 1999). We designed a series of laboratory experiments that asked people to reveal their preferences in a real auction in which they spent money and consumed the actual food products. We chose the lab approach to valuation after carefully considering and excusing the more standard methods. One alternative we considered was to use econometric techniques to tease out preferences from aggregated data collected for some other purpose. We decided that this method did not provide results we considered robust for our purpose since many of these new food products did not have a market to generate the needed data. We needed to create our own market. A second alternative was to conduct actual test-marketing of the food in a retail store. Apart from the obvious cost of this exercise we were also concerned we would lose control of the scientific setting. We needed to keep control of the many attributes of the goods, their quantity and quality, and the flow of information so we knew exactly what attributes the participants were valuing. The final alternative we shelved was to survey consumers in person by mail or by phone. The absence of a reasonable reality check in these surveys, however, caused some concern that participants might respond in unrealistic and biased manner. We wanted people to make real economic commitments, albeit in a setting more stylized than a retail store (also see (Hoffman et al. 1993)).

After ten years of work, these experimental procedures have passed a critical test. We have learned things about consumer behavior and welfare gains toward food safety that would have been impossible to discover from any of the alternative procedures we might have used. This paper describes some of what we learned about consumer attitudes about food safety, and about what insight can and cannot be learned in consumer experiments in the lab.

Food-borne Pathogens

Participants underestimate the objective risk of food-borne pathogens, but experience with the market and information on probabilities of illness and death influence their final assessment and valuation of these risks.

Evidence from laboratory auctions consistently suggests that people initially underestimate the risk of illness from food-borne pathogens (Hayes et al. 1995). In these auctions, participants indicated their willingness to pay to reduce the individual and combined risks of five different food-borne pathogens: *Campylobacter*, *Salmonella*, *Staphylococcus aureus*, *Trichinella spiralis*, and *Clostridium perfringens*.

Results generally indicate that people will pay significantly more for safer food after gaining auction experience and receiving the objective risk information. Figure 1 shows the average pre- and post-information bid by pathogen. People initially underestimated the risk associated with these pathogens, but adjusted upward their estimate after experience and objective information. Research suggests that marginal willingness to pay decreases as risk increases – indicating that people place more weight on their prior beliefs than on the objective information. While this general result appears to contradict the common finding that people overestimate the risks of low-probability events, the observation may be consistent with the broader interpretation that people underestimate extremely low levels of risk and overestimate less extreme low risks.

Participants seem to possess general preferences and values for food safety – rather than pathogen-specific preferences.

Since the risks from food-borne pathogens are relatively low compared to driving a car or other everyday activities, it is not completely surprising that the lab results suggest that people do not significantly differentiate pathogens when valuing food safety. In general, food-safety risks are relatively low on a daily basis and people might not distinguish between the risks posed by specific pathogens. If people do differentiate between specific pathogens, the values elicited for the combined risk from all pathogens should significantly differ from the values elicited for each individual pathogen. Results however suggest otherwise. Combined and pathogen-specific values were similar whether the person was acting on his or her own subjective perception of risk or on the objective risk level provided by experts (see Figure 1). The general values arising from the laboratory auctions indicate that the average participant was willing to pay approximately US$ 0.70 per meal for safer food. If one could transfer these values to the U.S. population, the value of food safety could be at least three times the largest previously available estimates. These participants had a significant demand for safer food, enough perhaps to justify the costs of current and future food-safety regulations.

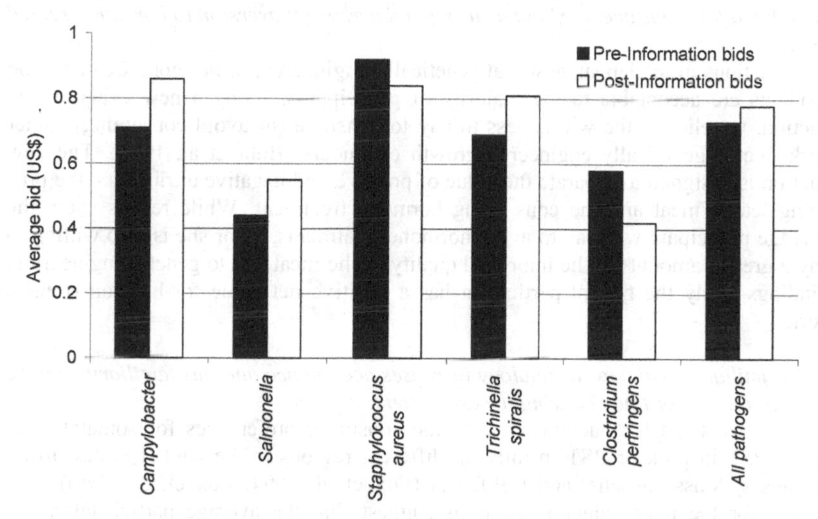

Figure 1. Average bid to exchange a risky sandwich for a less risky sandwich

Participants were willing to pay a price premium for new food products that they had not tried before.

But the US$ 0.70 food-safety premium exceeded some experts' expectations of what people would pay in retail markets. One explanation might be the novelty of the experimental experience. The open question is whether the act of bidding in a lab auction in a unique lab environment might have inflated the demand for food safety. Lab auctions are usually a one-time experience, and the concern is that people might experiment with their bids, bidding high because the costs of doing so are low. Theory, however, suggests an alternative explanation for the high price premia – the novelty of the food product. Many bidders have never experienced the goods up for auction, e.g., irradiated meat. In this case, theory says that a bid should reflect two elements of value – the consumption value of the good and the information value of learning how the good fits into his or her preference set. This idea of preference learning would exist if people bid large amounts for a good because they wanted to learn about an unfamiliar good they had not previously consumed, because it was unique, or because it was unavailable in local stores.

We tested these competing explanations by auctioning off three goods that vary in familiarity – candy bars, mangos, and irradiated pork, in four consecutive experimental auctions over two weeks. Their results suggest that preference learning, not novelty of the lab, seems to explain some of the price premia. No statistical change in bids was measured for candy bars and mangos, whereas the price premia for irradiated pork dropped by 50 percent over the four sessions. These findings suggest participants will pay a price premium for new products to learn how these goods might be an addition to their overall set of preferences. This suggests a premium of US$ 0.35 per meal for safer food, an amount that still exceeds previous estimates.

Growth Hormones

Participants generally preferred low calorie hormone-treated pork to typical food, but a few consumers exhibit a strong and persistent aversion to hormone-treated food.

Auctions in the lab suggest that genetically engineered, or hormone-treated, food products are acceptable to the majority of participants. Using a new experimental auction, we elicited the willingness to pay to consume (or avoid consuming) leaner pork due to genetically engineered growth enhancers (Buhr et al. 1993). The new auction is designed to separate the value of positive and negative attributes – the pros being leaner meat and the cons being hormone treatment. While results show the average participant will pay to avoid hormone treatments, he or she is also willing to pay a greater amount for the improved quality of the meat due to genetic engineering. Findings imply the typical participant has a positive net value for hormone-treated pork.

Familiarity with new technology increases acceptance and this familiarity can be learned locally, or taught during the experiment.

We used the lab auctions to examine consumer preferences for somatotropin, either PST in pork or BST in milk, in different regions in the United States: Iowa, Arkansas, Massachusetts, and California (Fox et al. 1994; Fox et al. 1995)). The results for the pork valuation auctions suggest that the average participant had a significant preference for the leaner pork yielded from the PST hormone treatment (Figure 2).

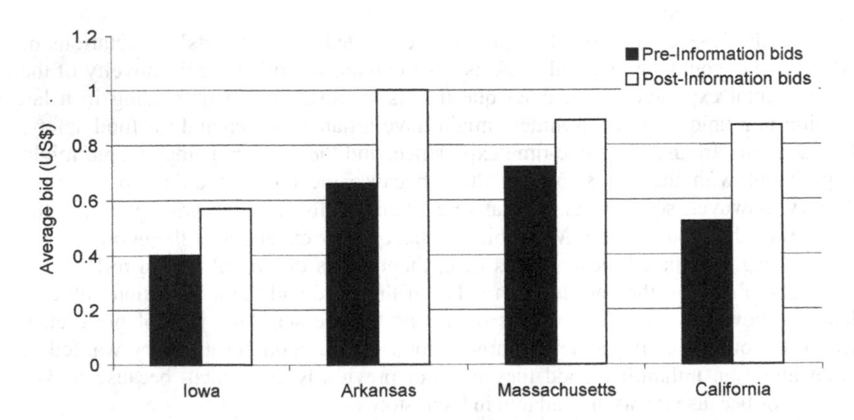

Figure 2. Average bid to exchange PST pork to NON-PST pork

We find similar results when eliciting consumer preferences for milk produced by cows treated with somatotropin. More than 60 percent of subjects indicated they would be willing to buy hormone-produced milk at little or no discount (see Figure 3). Two additional results emerge. First, preferences for hormone-treated products increased significantly as people became more informed about the treatment process. Second, rural Californians were very familiar with the technology and were willing to try it. Urban Californians, however, knew little about the technology, but they quickly accepted and valued the process once it was explained.

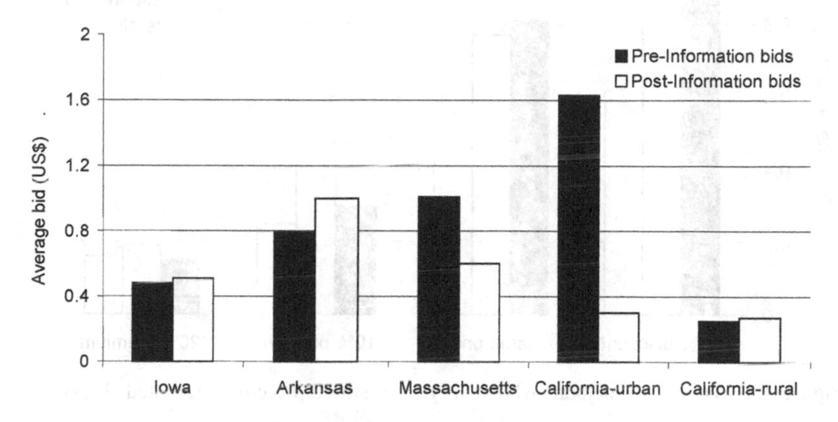

Figure 3. Average bid to exchange BST milk for NON-BST milk

Irradiation

Irradiation appears to be acceptable. Most participants were willing to pay a premium for irradiated food. Laboratory auctions indicate that participants are not averse to using irradiation as a risk-reduction technology.

For a comparative baseline, we used the lab auctions to elicit participant willingness to pay for safer chicken breasts without disclosing the risk-reduction technology. We then compared these baseline results to equivalent auctions in which the technology was disclosed to be irradiation with standard USDA information. Consumer willingness to pay was statistically equal in each case – approximately US$ 0.80 per chicken breast. We also observed that nearly 80 percent of the laboratory consumers preferred the irradiated chicken to the non-irradiated chicken if it was available for the same price (Shogren et al. 1999). Thirty percent of the consumers were willing to pay a 10-percent premium for the irradiated chicken, and twenty percent were willing to pay a 20-percent premium (Figure 4). Results, therefore, strongly suggest that irradiation is an acceptable risk-reduction technology to informed consumers and estimates of willingness to pay for irradiation more than covers the cost of commercial-scale implementation.

Negative reports concerning irradiation had a larger impact on participant preference and values than positive reports – even when the negative reports were unscientific.

Some of the results we found were puzzling. Our participants in the lab appear to be very accepting of new technologies, whereas the average American is not. The key to this conundrum is that our experimental design controlled the flow of information about irradiation, and in most cases, our formal descriptions of the new technology suggested that the process was safe and beneficial. The lab allowed us to address this issue directly, and one of our most surprising results came about when we experimented with negative descriptions taken from activist groups.

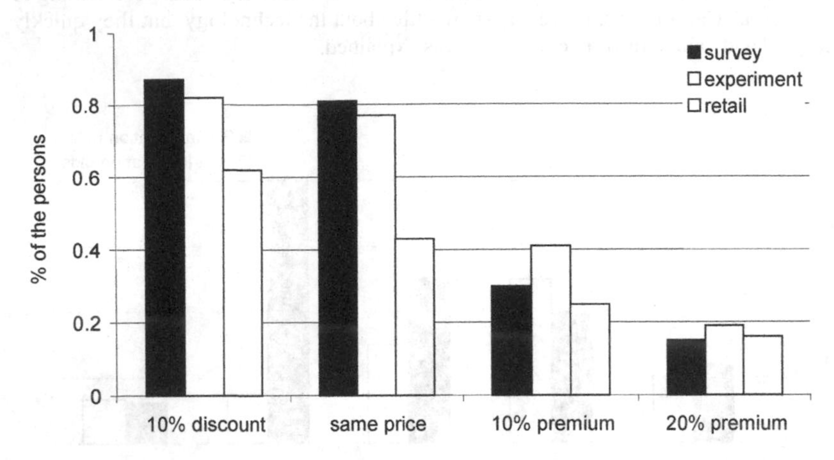

Figure 4. Percentage of persons who said 'yes' to stated price for irradiated chicken

In this set of auctions, we examined how consumer willingness to pay for safer pork sandwiches was affected by alternative descriptions of food irradiation (Fox, Hayes and Shogren 2002). Results follow intuition with favorable description of irradiation increasing willingness to pay and unfavorable descriptions decreasing willingness to pay. But when presented with both a favorable and an unfavorable description, the participants acted as if they had read only the negative information – indicating that the negative portrayal dominated the positive (see Figure 5).

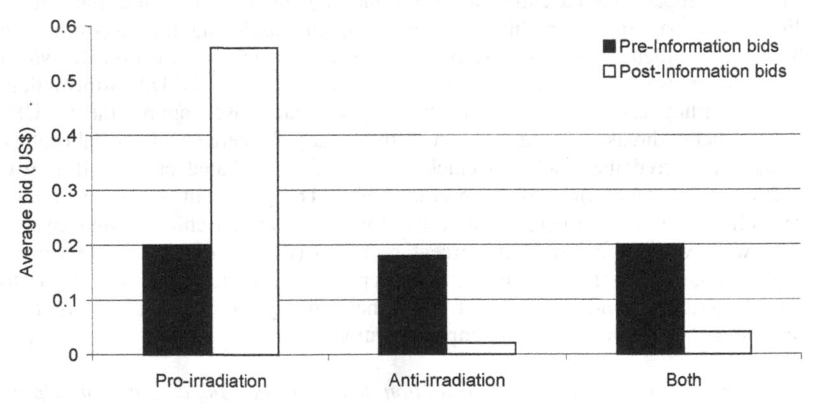

Figure 5. Average bid to exchange meat for irradiated meat before and after information

This relative impact of the unfavorable description was evident even when the negative representation was a non-scientific account written by a consumer advocacy group. This result illustrates the incentive that partisan groups have to promote unscientific claims to advance an agenda that yields possible loss in general social welfare. It is always possible to describe a new food process in a way that suggests that it is unsafe. For example, one can make the statement that "scientists cannot be 100% sure this food does not cause cancer" about any food or food process. Our experimental work convinced us that when the media give equal billing to those who are prepared to make this kind of statement, public opinion can quickly turn against the food or food process. This result has immediate relevance to the ongoing debate about genetically modified foods. Negative information dominates, and it then becomes a question of whether a neutral third party exists to supply verifiable information written by concerned but neutral interests.

Concluding remarks

Consumers value safer food. Understanding how they value less risk from food requires tools that can isolate food quality and quantity from food-borne risks. We have reviewed how experimental methods can be used as a tool to isolate and control the market setting to address specific questions on how people value new and controversial food products. One example stands out – when faced with both positive and negative information about new food technologies, participants reacted as if they had received only negative information. They seemed to react to only the bad news, irrespective of the source. We have also learned that limits exist to what can be achieved with lab experiments for valuation work. We had hoped to collect refined information about the value of reductions in individual pathogens, but we discovered that we could detect only general preferences about food safety. We found that subtle changes in the experimental procedure such as whether we paid the participants ahead of time, the choice of auction, asked for willingness to pay or willingness to accept, or posted market-clearing prices could significantly impact the results. Finally, we discovered that bids for new foods or food processes could be unrealistically high when participants viewed them as a novelty. But despite these limits, our experience leads us to conclude that over time as designs are refined, improved reality-based consumer experiments will become an increasingly important method for applied economists interested in the demand side of food safety.

References

Brundtland, G., 2001. *How safe is our food? Statement by the director-General.* Available: [http://www.who.int/fsf/dgstate.htm] (6 Mar 2003).

Buhr, B. L., Hayes, D. J., Shogren, J. F., et al., 1993. Valuing ambiguity: the case of genetically engineered growth enhancers. *Journal of Agriculture and Resource Economics,* 18 (2), 175-184.

Buzby, J. and Roberts, T., 1996. Microbial foodborne illness: the costs of being sick and the benefits of new prevention policy. *Choices,* 11 (1), 14-17.

Buzby, J., Roberts, T., Lin, C. T. J., et al., 1999. *Bacterial foodborne disease: medical costs and productivity losses.* Economic Research Service, USDA, Washington D.C. Agricultural Economic Report no. 741.

Caswell, J., 1995. *Valuing food safety and nutrition.* Westview Press, Boulder, CO.

Crutchfield, S. R., Buzby, J. C., Roberts, T., et al., 1997. *An economic assessment of food safety regulations: the new approach to meat and poultry inspection.* Economic Research Service-USDA, Washington, DC. Agricultural Economic Report no. 755.

Food safety - a worldwide public health issue, 2000. Available: [http://www.who.int/fsf/fctshtfs.htm] (6 Mar 2003).

Food safety strategic plan, 2001. Available: [http://www.foodsafety.gov/~fsg/cstrpl-4.html] (6 Mar 2003).

Fox, J., Buhr, B., Shogren, J., et al., 1995. A comparison of preferences for pork sandwiches produced from animals with and without somatotropin administration. *Journal of Animal Science,* 73 (4), 1048-1054.

Fox, J., Hayes, D. and Shogren, J., 2002. Consumer preferences for food irradiation: how favorable and unfavorable descriptions affect preferences for irradiated pork in experimental auctions. *Journal of Risk and Uncertainty,* 24, 75-95.

Fox, J. A., Hayes, D. J., Kliebenstein, J. B., et al., 1994. Consumer acceptibility of milk from cows treated with bovine somatotropin. *Journal of Dairy Science,* 77 (3), 703-707.

Freeman, A. M., 1993. *The measurement of environmental and resource values: theory and methods*. Resources for the Future, Washington, DC.

Hayes, D. J., Shogren, J. F., Shin, S. Y., et al., 1995. Valuing food safety in experimental auction markets. *American Journal of Agricultural Economics,* 77 (1), 40-53.

Hoffman, E., Menkhaus, D., Chakravarti, D., et al., 1993. Using laboratory experimental auctions in marketing research: a case study of new packaging of fresh beef. *Marketing Science,* 12, 318-338.

Kaferstein, F. and Abdussalam, M., 1999. Food safety in the 21st century. *Bulletin of the World Health Organization,* 77 (4), 347-351.

Kuchler, F. and Golan, E., 1999. *Assigning values to life: comparing methods for valuing health risk.* Food and Rural Economics Division, Economics Research Service, US Department of Agriculture, Washington. Agricultural Economic Report no. 784.

Lutter, R., 1999. Food irradiation--the neglected solution to food-borne illness. *Science,* 286 (5448), 2275-2276.

Mead, P. S., Slutsker, L., Dietz, V., et al., 1999. Food-related illness and death in the United States. *Emerging Infectious Diseases,* 5 (5), 607-625. [http://www.cdc.gov/ncidod/eid/vol5no5/mead.htm]

Miller, M. and Altekruse, S., 1998. The president's national food safety initiative. *Journal of the American Veterinary Medical Association,* 213 (12), 1737-1739.

Schelling, T., 1984. The life you safe may be your own. *In:* Schelling, T. C. ed. *Choice and consequence : perspectives of an errant economist.* Harvard University Press, Cambridge, MA., 113-146.

Shogren, J. F., Fox, J. A., Hayes, D. J., et al., 1999. Observed choices for food safety in retail, survey, and auction markets. *American Journal of Agricultural Economics,* 81 (5), 1192-1204.

Shogren, J. F., Shin, S. Y., Hayes, D. J., et al., 1994. Resolving differences in willingness to pay and willingness to accept. *American Economic Review,* 84 (1), 255-270.

Unnevehr, L. J. and Jensen, H. H., 1999. The economic implications of using HACCP as a food safety regulatory standard. *Food Policy,* 24 (6), 625-635.

Viscusi, W. K., 1992. *Fatal tradeoffs: public & private responsibilities for risk.* Oxford University Press, Oxford.

3

Consumer perception of food safety: role and influencing factors

Wim Verbeke[*]

Introduction

Food production and consumption have been under heavy criticism during the last decade. Many organizations including consumers, industry, producers and governments, as well as scientists from a plethora of disciplines, have recently been involved in debates that were initiated by numerous food-safety crises. From all debated food items, meat is referred to as the food item in which consumer confidence decreased most during the last decade (Richardson, MacFie and Shepherd 1994; Issanchou 1996; Becker, Benner and Glitsch 1998). Therefore, research into consumer decision-making towards fresh-meat consumption is chosen as the showcase to discuss food-safety issues from the perspective of the demand side of the food chain, i.e., the consumer.

The relevance of meat issues and a better understanding of consumer decision-making towards meat became paramount due to distinct changes at the consumer level. Along with increasing importance of quality, organoleptic and sensory properties of food, issues relating to food safety, human health and wellbeing have gained attention, especially with respect to fresh-meat production and consumption. Meat has traditionally constituted a substantial part of the West-European diet. Increasing economic and social welfare since the 1950s resulted in increasing amounts of animal-protein intake. Top meat consumption levels were noticed during the first half of the nineties in most of the EU countries, but ever since, fresh-meat consumption levels decreased.

This paper focuses on potential contributions from behavioral sciences (consumer behavior and marketing) to issues related to food safety. The objectives of this paper are to review insights in consumer decision-making towards fresh-meat consumption in situations with uncertainty and risk. This paper summarizes findings from empirical research implemented in Belgium during the period 1996-2002, which were fully presented in several publications. Readers are referred to the original publications for the details of the methodological approaches, empirical analyses, tables and graphs.

Economic impact assessment

The economic impact of food-safety crises is hard to assess. If, for example, the BSE crisis is considered, the direct cost of the obligatory BSE-testing scheme for bovine animals aged beyond 30 months, which has been operational since January 2001, is estimated at € 100 per test. For the specific case of Belgium, it is estimated that about 300,000 of those tests are to be performed every year, which yields a total

[*] *Department of Agricultural Economics, Ghent University, Coupure links 653, B-9000 Gent*

A.G. J. Velthuis et al. (eds), New Approaches to Food-Safety Economics, 21-26.

direct cost of € 30 million per year. Furthermore, as a result of several EU feed bans (e.g. the bans on feed containing meat-and-bone meal and serious-risk material), some 600,000 tons of hazardous material has to be destroyed every year. This can be realized at a cost of about € 0.15 – € 0.20 per kg, leading to a total cost of about € 100 million per year.

To pass the direct testing costs and the costs resulting from the new legislation on to the consumer, meat prices would have to rise by 1.25% to 1.75% at the retail level (Verbeke 2001b). Combined with an inelastic price-elasticity coefficient for beef of approximately -0.5 (Verbeke and Ward 2001), such a price increase would result in a decrease of beef demand by less than 1%. This decrease does not match with the actual 4-5% decrease in beef demand, which the industry has faced every year since the BSE crisis. This case exemplifies the decreasing power of neo-classical economic theory to explain contemporary changes in consumer behavior, as has already been shown by Bansback (Bansback 1995). When analysing factors influencing meat demand, Bansback (Bansback 1995) concluded that economic factors explained 60% of the changes in meat demand during the period 1975-1994, while that share amounted to almost 90% for the period 1955-1974. Emerging factors include changing taste and preference patterns and consumer confidence. The issue of consumer confidence should also be considered as a major cost item, though one that is extremely difficult to quantify in economic terms. Studies of consumer behavior from a sociological or marketing perspective can shed light on the role and importance of consumer confidence related to food, safety and health.

Meat consumption behavior

The evolution of meat consumption in Belgium since 1955 reveals that distinct long-term changes have taken place. Animal-protein and fat intake have risen along with increasing wealth in the West-European society. Over time, a gradual shift away from red to white meat types was observed. Top meat-consumption levels were reached during the first half of the nineties, with considerable consumption decreases being noticed since, especially when considering per capita at-home meat intake. Over a period of seven years (1995-2001) Belgian at-home fresh beef and veal consumption fell more than 30%, while pork and poultry consumption decreased by about 8% and 5%, respectively. Out-of-home meat consumption may have increased but figures are not readily available. Nevertheless, available data from supply balance sheets and household panels systematically point towards recent significant consumption declines, which exemplifies a general "malaise" against fresh meat.

With respect to fresh-meat consumption frequency, it was found that daily fresh-meat consumers were the least inclined to reduce their consumption level as compared to less frequent consumers. Heavy meat consumers showed the strongest intentions to maintain their consumption levels, while less frequent consumers intended to decrease their fresh-meat consumption frequency further, herewith moving away from consuming fresh meat daily to several times a week or a lower frequency (Verbeke, Ward and Viaene 2000). Furthermore, the group of low-frequent meat consumers who show the strongest intentions to cut consumption further is gradually growing over time.

Consumer attitude: perception of fresh meat

In the consumer psychology and behavior disciplines, it is widely recognized that there exists a distinct filter or gap between the external (objective) and the internal (subjective) world of consumers (Risvik 2001). This filter, also called a perception filter, accounts for the difference between scientific objectivity and human subjectivity. The paramount importance of human subjectivity or perception lies in the fact that exactly perception – and not necessarily scientific facts – determines preference and choice. Therefore, this perception should be of interest to health and nutrition policymakers, and is definitely of interest to the food industry.

The importance of consumer perception has been assessed during two consumer surveys (1998 and 2000). Attribute-rating profiles of April 1998 reveal that problems of the beef image pertained to safety and trustworthiness. Pork was characterized as the most fat, the worst tasting, the least healthy and the overall lowest-quality meat (Verbeke and Viaene 1999a). Poultry received the best overall perception scores.

The same measurement was repeated two years later, in April 2000, after the occurrence of the Belgian dioxin crisis. Like the previous meat-safety crises (hormone abuses, antibiotic residues, BSE), the dioxin scare received considerable attention from the mass media, which brought the issue to the public's attention in May 1999. Pork and especially poultry were affected by the dioxin crisis, which was clearly reflected in their perceptual profiles. This led to significant shifts towards the "with hormones" pole of the semantic differential scale (or stronger associations with containing potentially harmful substances) for the perception of both meats types. Additionally, perception of poultry on "quality", "trustworthiness" and "safety" significantly worsened after the dioxin crisis. No other shifts of the pork and poultry perception profiles were formed, which is reasonable in the absence of substantial changes in sensory, price, convenience or animal-welfare issues during the considered time interval (Verbeke 2001a). Reversibly, beef perception improved on safety attributes. Remarkably, beef consumption continued to decrease whereas pork and poultry consumption stabilized. Furthermore, about 25-30% of the consumers reported high concerns about BSE in poultry or dioxins in beef, which is in direct contrast with scientific evidence.

Considerable bias was discovered between meat facts or scientific-indicator criteria and consumer perception of these facts. This phenomenon has specifically been addressed related to health, leanness and sensory characteristics of pork (Verbeke et al. 1999), but was also related to meat-quality labels (Verbeke and Viaene 1999b). Pork perception was found to be worst as compared to beef and poultry on "leanness", "healthiness", and attributes that relate to eating or sensory quality, i.e. "taste" and "tenderness". On the contrary, it was scientifically shown that pork can be low in fat and cholesterol, or excelling in taste and tenderness, depending on the specific cut and handling throughout the meat chain. Similar conclusions were drawn related to the perception of quality labels. A considerable part of the interviewed consumers claimed to buy labeled meat but failed to recall any label unaided. Additionally, features and benefits are assigned to quality-labeled meat that do not correspond to the actual performance of the label.

Impact of communication

The gap between scientific facts and their perception by consumers is largely shaped by communication. Claimed attention to mass-media publicity was found to have a strongly negative influence on consumer behavior and decision-making processes towards fresh meat. Consumers, who attended mass-media coverage of fresh-meat issues, reported significantly higher meat-consumption decreases with reference to the past as well as stronger intentions for decrease in the future. It was also found that consumers who pay a high level of attention to media reports, express higher health consciousness, more misperception of health risks and higher levels of concern about potential health hazards that were frequently reported in mass media. While the impact of attention to mass-media publicity was shown to be very significant, high levels of attention to personal communication from butchers or to advertising were found to have some but far more limited impact. Meat consumers who pay high levels of attention to information from butchers reported more positive meat-attribute perception scores, but this did not translate into associations with health concern, claimed behavior or behavioral intention (Verbeke, Viaene and Guiot 1999).

The negative impact of television publicity was confirmed through econometric cross-sectional and time-series analyses. Probabilities to cut fresh-meat consumption were boosted as consumers reported to have paid high attention to television coverage of meat issues (Verbeke, Ward and Viaene 2000). Similarly, parameters of television coverage indices were largely significant and negative in an Almost Ideal Demand System for fresh meat, contrary to the estimates of the advertising-expenditure variables, which were insignificant. In case of beef in Belgium during the second half of the nineties, a negative press over advertising impact ratio of five to one was found (Verbeke and Ward 2001). It means that five units of positive news are needed to offset the impact of one similar negative message.

Potential from labeling and traceability

It has been indicated that a label can serve as an important extrinsic product-quality cue in the evaluation process (Caswell 1992; Issanchou 1996). Furthermore, meat labeling has been reported as a promising strategy to regain consumer confidence (Wagner and Beimdick 1997; Wit et al. 1998). In line with these empirical findings, the Commission of the European Communities enacted regulations concerning the establishment of a system of identification and registration of bovine animals, as well as the compulsory labeling of fresh and frozen beef and beef products. This system has been fully operational since January 2002 and makes beef labeling possible, including a traceability reference number, slaughterhouse and cutting-unit license numbers, and name of country in which the animal was born, raised and slaughtered. Despite numerous labeling efforts by industry and government, knowledge and perception of labels were discovered to contrast with exact labeled-product features, but over time the situation seems to improve. Consumers who experienced (bought) meat with a quality label reported a more favorable attitude towards and a better knowledge of labeled meat (Verbeke and Viaene 1999b).

The rational support that consumers seek when making meat-purchasing decisions can be delivered through the establishment of a waterproof system of identification, traceability and control, eventually sealed with a label for recognition and additional assurance. Traceability, labeling and assurance schemes are established to allay consumer concerns, but the standards and realizations are difficult to communicate and at risk of being perceived by consumers as insufficient or meaningless. Intrinsic opportunities such as the ability to organize the chain more efficiently, monitor the chain, and assess individual responsibilities, are broadly supported by consumers and therefore also an issue of public policy and regulation. Extensions with respect to process attributes, such as alternative production methods, origin and labeling, are less relevant to the broad public and only of interest to specific market segments (niches). Therefore, intervention on the process attribute side is most appropriate for private initiatives and embraces opportunities for product differentiation and competitive advantage in well-defined markets (Gellynck and Verbeke 2001; Verbeke 2001c).

Conclusions

Several years of consumer research on meat consumption in Belgium yielded a comprehensive picture of the impact of meat-safety issues. There is no doubt about the existence of considerable misperception by consumers, lack of knowledge and bias between perception and scientific-indicator criteria related to health and safety characteristics of meat. Generally, mass-media publicity – following problems or abuses throughout the food chain – is found to have significantly and negatively affected decision-making on fresh-meat consumption. On the contrary, little evidence of effects of positive communication could be shown.

As the starting point of the meat chain, livestock farming urgently needs to be reoriented towards quality in a broad sense instead of quantity-oriented mass production. This includes product and process quality, animal welfare and environment preservation. Quality, health, convenience and variety-seeking trends at the consumer level urge for product and process innovation, as well as for the adoption of new technologies and quality control at the meat-industry level. Finally, the role of the government is twofold. First, it consists of protecting the consumers through guarding the information spreading and through providing extension related to potential health risks and benefits. Second, it pertains to provide clear and unambiguous legal frameworks, including both the establishment and control of production and product standards. The establishment of regulated traceability systems is a big leap forward, although it remains an issue of debate whether something is gained in terms of intrinsic food safety. Definitely, traceability yields opportunities to gain in terms of consumer confidence and food-safety perception.

Taking away any basis for negative press should be a priority for the food sector. This can be realized through producing safe and sound products, through acceptable production methods with respect to animal welfare and the environment, and through practical applications of comprehensive quality and chain-monitoring systems. Finally, correct products have to be accompanied by trustworthy communication. This communication challenge is probably harder to realize than the product-safety challenge, but the reward in terms of consumer trust and product acceptance is definitely worth striving for.

References

Bansback, B., 1995. Towards a broader understanding of meat demand. *Journal of Agricultural Economics,* 46 (3), 287-308.

Becker, T., Benner, E. and Glitsch, K., 1998. *Summary report on consumer behaviour towards meat in Germany, Ireland, Italy, Spain, Sweden and The United Kingdom - results of a consumer survey.* Universität Hohenheim, Göttingen. Working Paper FAIR CT-95-0046.

Caswell, J., 1992. Current information levels on food labels. *American Journal of Agricultural Economics,* 74 (5), 1196-201.

Gellynck, X. and Verbeke, W., 2001. Consumer perception of traceability in the meat chain. *Agrarwirtschaft,* 50 (6), 368-374.

Issanchou, S., 1996. Consumer expectations and persceptions of meat and meat product quality. *Meat Science,* 43, S5-S19.

Richardson, N., MacFie, H. and Shepherd, R., 1994. Consumer attitudes to meat eating. *Meat Science,* 36 (1/2), 57-65.

Risvik, E., 2001. The food and I: sensory perception as revealed by multivariate methods. *In:* Frewer, L., Risvik, E. and Schifferstein, H. eds. *Food, people and sociey: a European perspective of consumers' food choices.* Springer-Verlag, Heidelberg, 23-37.

Verbeke, W., 2001a. Beliefs, attitude and behaviour towards fresh meat revisited after the Belgian dioxin crisis. *Food Quality and Preference,* 12 (8), 489-498.

Verbeke, W., 2001b. Consumer reactions and economic consequences of the BSE crisis. *Verhandelingen van de Koninklijke Academie voor Geneeskunde van België,* 63 (5), 483-492.

Verbeke, W., 2001c. The emerging role of traceability and information in demand-oriented livestock production. *Outlook on Agriculture,* 30 (4), 249-255.

Verbeke, W., Van Oeckel, M., Warnants, N., et al., 1999. Consumer perception, facts and possibilities to improve acceptability of health and sensory characteristics of pork. *Meat Science,* 53 (2), 77-99.

Verbeke, W. and Viaene, J., 1999a. Beliefs, attitude and behaviour towards fresh meat consumption in Belgium: empirical evidence from a consumer survey. *Food Quality and Preference,* 10 (6), 437-445.

Verbeke, W. and Viaene, J., 1999b. Consumer attitude to beef quality labeling and associations with beef quality labels. *Journal of International Food and Agribusiness Marketing,* 10 (3), 45-65.

Verbeke, W., Viaene, J. and Guiot, O., 1999. Health communication and consumer behavior on meat in Belgium: from BSE until dioxin. *Journal of Health Communication,* 4 (4), 345-357.

Verbeke, W. and Ward, R., 2001. A fresh meat almost ideal demand system incorporating negative TV press and advertising impact. *Agricultural Economics,* 25 (2/3), 359-374.

Verbeke, W., Ward, R. and Viaene, J., 2000. Probit analysis of fresh meat consumption in Belgium: exploring BSE and television communication impact. *Agribusiness,* 16 (2), 215-234.

Wagner, P. and Beimdick, E., 1997. Determinanten des Erfolgs von Markenfleischprogrammen. *Berichte über Landwirtschaft,* 75 (2), 171-205.

Wit, M. A. d., Koopmans, M. P., Kortbeek, L. M., et al., 1998. Consumer-oriented new product development: principles and practice. *In:* Meulenberg, M. ed. *Innovation of food production systems : product quality and consumer acceptance.* Wageningen Pers, Wageningen, 37-66.

4

Quantitative risk assessment of food borne pathogens – a modeling approach

E.G. Evers, M.J. Nauta, A.H. Havelaar and A.M. Henken[*]

Introduction

There is always a certain probability that the consumption of food leads to a reduced health status due to microbial contamination, however small this risk may be. One of the most frequently occurring effects of food infection or food poisoning is the occurrence of gastroenteritis. Usually this means an acute and temporary reduction in health status at the individual level, although also chronic effects and even death may occur. In this contribution we would like to describe:

- how frequently gastroenteritis occurs and which micro-organisms are involved;
- in what way we may get insight into means to control or, preferably, reduce health damage due to food-borne pathogens.

Quantitative microbiological risk assessment will be introduced. More specifically we will present work on modeling the fate of pathogens along the production chain, dose response and disease burden.

How frequently does gastroenteritis occur?

To obtain an estimate on how many cases of gastroenteritis occur in the Netherlands, data from several monitoring and registration systems are available. However, each system will lead to a different estimate, because mostly only a certain part of the human population is investigated (Figure 1). Using a population-based cohort study would allow for an estimate of the total number of patients involved. In the early nineties such a population study was performed in four regions in the Netherlands (Hoogenboom-Verdegaal et al. 1994) and more recently, in 1999, a second population study was done (De Wit et al. 2001b). The gastroenteritis incidence was estimated at 283 per 1,000 person-years in this last study (about 4.5 million cases per year in the Dutch population of 15.6 million).

A second way to obtain an estimate of the number of patients is through monitoring how many patients consult their general practitioner with gastroenteritis complaints using sentinel general practices (GP). The Netherlands Institute for Health Services Research maintains a network of sentinel general practices that participate in a continuous morbidity registration. Recent results show that about 220,000 persons consult their general practitioner each year for gastroenteritis, which corresponds to about 5% of people with gastroenteritis (De Wit et al. 2001a; De Wit 2002).

---

[*] *Microbiological laboratory for health protection, National Institute for Public Health and the Environment, P.O. Box 1, 3720 BA Bilthoven, The Netherlands*

A.G. J. Velthuis et al. (eds), New Approaches to Food-Safety Economics, 27-37.

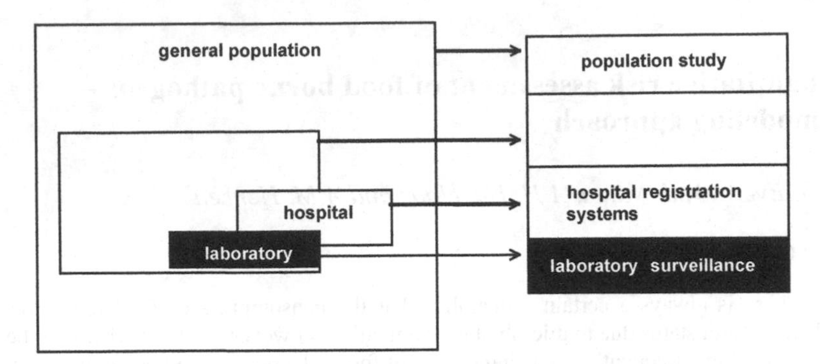

Figure 1. The various systems to obtain data on frequency of gastroenteritis.

In addition to only a minority of patients with gastroenteritis seeking medical care, only in a minority of cases the attending physician requests for microbiological testing, of which only a proportion leads to a positive result. In the Netherlands a laboratory-based surveillance system exists for some bacterial pathogens. This is a continuous system, suited to investigate trends. Other existing registration systems are: registration of hospital discharge diagnosis, mortality registration, statutory notifications, registration of outbreaks through the Municipal Health Services and through the Inspectorate for Health Protection and Veterinary Public Health.

In the recently performed sentinel GP study, De Wit et al. (De Wit et al. 2001a) found that *Campylobacter* could be detected most frequently (10% of cases), followed by *Giardia lamblia* (5%), rotavirus (5%), Norwalk-like viruses (5%) and *Salmonella* (4%). A pathogen could be found in almost 40% of the gastroenteritis patients. The ranking of the various micro-organisms found was different in the population-based study (De Wit et al. 2001b), where viral pathogens were the most prominent pathogens, detected in 21% of cases, with Norwalk-like viruses as the most frequent ones (11%).

The total number of salmonellosis cases as estimated in the population study is about 3 per 1,000 person-years, indicating a number of about 50,000 cases annually in the Netherlands (De Wit et al. 2001b). The total number of laboratory-confirmed salmonellosis cases in the Netherlands has since long been decreasing (Van Pelt et al. 1999). The incidence of *Campylobacter*-associated gastroenteritis is estimated at 6.8 per 1,000 person-years, indicating a number of about 100,000 cases annually in the Netherlands (De Wit et al. 2001b).

The food-attributable fraction in campylobacteriosis cases is estimated at > 90% (*Voedselinfecties* 2000). From a review (Altekruse et al. 1999) it can be estimated, mainly based on case-control research, that 70-90% of the *Campylobacter* infections are associated directly or indirectly with poultry meat. However, more recently there are indications that other transmission routes may be more important than originally thought (Havelaar et al. 2000). During the dioxin crisis in Belgium in 1999, when Belgian poultry was withdrawn from the market, the number of campylobacteriosis cases was reduced by 40% (Vellinga and Van Loock 2002).

The number of pathogens that may infect people through food is large and the same is true for the number of food items in which pathogens can be found (De Boer 2000). As resources are limited and information scarce the Inspectorate for Public Health Protection and Veterinary Public Health needed tools how to select the most important micro-organisms among the ones present. Research aimed at prevention of existing and (re-)emerging problems could then be more focussed than hitherto. To this end a discussion group was formed that came up with criteria and weighing methods (Themarapport Gezonde Voeding & Veilig Voedsel, in preparation). The following five selection criteria were used: frequency of occurrence; severity when occurring; chance for explosions; is the disease endemic or can it become endemic; and whether or not there are specific research needs.

Although information on food-borne infections is becoming available as indicated above it is still scarce (*Voedselinfecties* 2000). Risk analysis is advocated as a means to come to objectives for food-safety control.

Risk analysis and risk assessment

Risk analysis consists of three coherent activities: risk assessment, risk management, and risk communication (*Proposed draft principles and guidelines for the conduct of microbiological risk assessment* 2001). Risk assessment is a scientifically based process, consisting of Hazard Identification (identification of the agent causing adverse health effects), Exposure Assessment (evaluation of the intake of the agent), Hazard Characterization (evaluation of the nature of the adverse health effects), and Risk Characterization (estimation of occurrence and severity of the adverse health effects) (*Codex Alimentarius Commission. Appendix II: Draft principles and guidelines for the conduct of microbiological risk assesment* 1998).

The functional separation of risk assessment from risk management helps assure that the risk-assessment process is unbiased. However, certain interactions are needed for a comprehensive and systematic risk-assessment process. The benefits of the use of risk assessment, or more specifically of quantitative microbiological risk assessment (QMRA), is in our view threefold:

- it results in an estimate of the health risk of a certain pathogen / product / population combination. This provides an alternative to epidemiological research;
- it is possible to make comparisons of the relative importance, in terms of public health, of e.g. different pathogens in a product, or of a certain pathogen in different products; and
- most importantly, it provides estimates of the effect of a certain intervention, again in terms of public health.

We implement risk assessment by starting with an exposure-assessment model, which has an exposure estimate as an output (e.g. the number of pathogens per serving and the probability of contaminated servings combined with the number of servings per day). This is then the input of a hazard-characterization model which converts the exposure into an estimate of the public health risk of the considered pathogen / product / population combination.

At the beginning of a risk-assessment project, the specific purpose of the particular risk assessment being carried out should be clearly stated as the statement of purpose. The output form and possible output alternatives of the risk assessment should be defined. In addition, the statement of purpose should usually also contain detailed demarcations to obtain a realistic project size (Nauta et al. 2001). These are:

- product definition: exactly which product (including details on production and processing) is considered;
- species/serotype definition: exactly which type or set of types is to be considered; and

– interventions: which interventions are to be considered, as the mathematical model to be developed must be able to include these interventions.

Exposure assessment

A first example of exposure-assessment modeling is a model for the transmission of *Salmonella* through the poultry meat production chain (Nauta, Van De Giessen and Henken 2000). The model first describes the situation before intervention (1997) in terms of *Salmonella* prevalences at flock level and some transmission parameters. The model input parameters were derived from expert opinion as research data were lacking. The effects of two intervention strategies for the Dutch poultry industry were predicted.

A general framework for performing exposure assessment, the Modular Process Risk Model (MPRM), was recently proposed (Nauta et al. 2001; *Risk assessment of food borne bacterial pathogens: quantitative methodology relevant for human exposure assessment* 2003). At the heart of the proposal is the suggestion that to each of the steps or key activities at the various intermediary stages of a farm-to-fork chain at least one of six basic processes can be assigned. These basic processes are the six fundamental events that may affect the transmission of any microbial hazard in any food process. There are two 'microbial' basic processes, growth and inactivation, and four 'food handling' processes, mixing and partitioning of the food matrix, removal of a part of the units, and cross-contamination.

In microbial risk assessment, calculations with numbers (N) of micro-organisms are to be preferred to concentrations (C). The rationale of using N instead of C in the calculations is that one is forced to do realistic calculations with discrete numbers, which is particularly relevant when N is small. For each step in the food pathway we are interested in the input-output relation for the number of cells per unit of product, *N*, the fraction of contaminated units (the prevalence), *P*, and the unit (Figure 2). The unit is a physically separated quantity of product in the process, for example an animal or a bottle of milk. Units might have to be redefined for each stage.

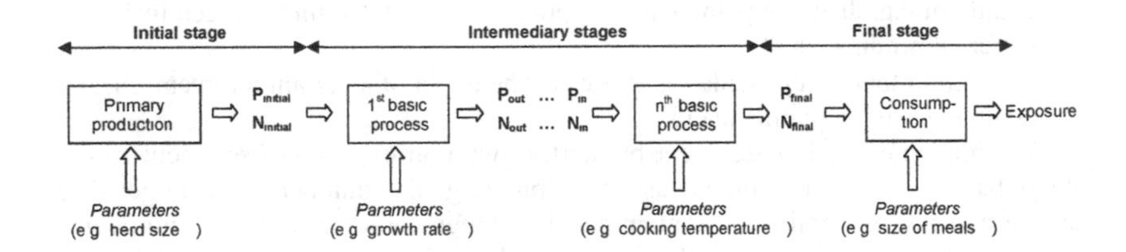

Figure 2. Schematic representation of a food pathway split up into different steps, each represented by an input-output basic process. P and N are prevalence and level of contamination, respectively (*Risk assessment of food borne bacterial pathogens: quantitative methodology relevant for human exposure assessment 2002*; modified).

The MPRM approach presented above was applied to Shiga toxin-producing *Escherichia coli* O157 (STEC O157) in steak tartare patties (see Figure 3) (Nauta et al. 2001). As slaughter practices may differ, three routes of exposure were compared, separating 'industrial' and 'traditional' ways of both slaughter and subsequent processing. Also, three preparation styles of the steak tartare patties (raw, medium and well done) were considered. As for a large part of the model parameters the information required to estimate their values was lacking, an expert elicitation workshop was organized. The model was implemented in @Risk (an Add-In of Excel) and analysed using Monte Carlo simulations.

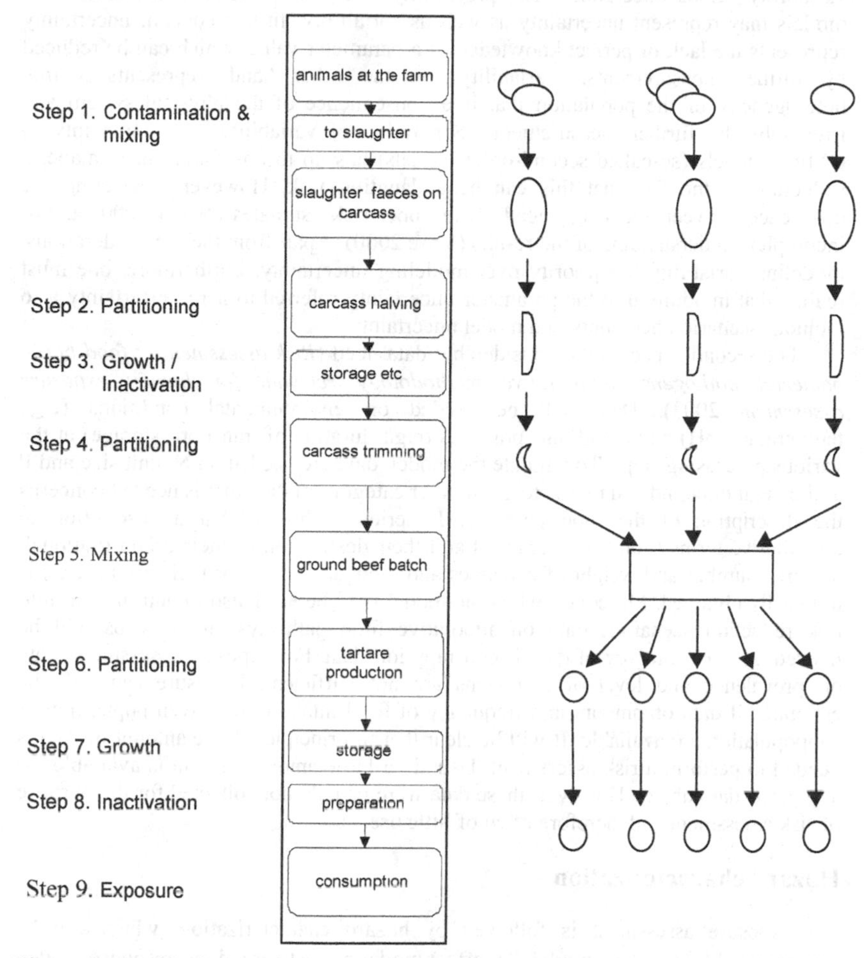

Figure 3. Food pathway of steak tartare. The pathway is split up into nine modeling steps. A basic process is assigned to each step. The illustration shows how the units are formed by partitioning and mixing.

The exposure model predicts that about 0.3% of the raw steak tartare patties is contaminated with STEC O157. Of these contaminated patties, a large fraction (>60%) is contaminated with one colony-forming unit (cfu) only. High contamination levels are rare, with for example only 7% of the contaminated raw steak tartare patties containing more than 10 cfu. As a comparison, in a microbiological survey it was found that one out of 82 raw steak tartare patties (1.2%) was positive for STEC O157. Knowing that the probability of detection of single cfus in such a survey is small, this suggests that the model prediction is an underestimation of the actual level of contamination of steak tartare patties.

Two important general aspects related to QMRA will be considered. The first is variability versus uncertainty. The probability distributions used in stochastic risk models may represent uncertainty as well as variability. In this context, uncertainty represents the lack of perfect knowledge of a parameter value, which can be reduced by further measurements. Variability, on the other hand, represents a true heterogeneity of the population that is a consequence of the physical system and irreducible by further measurements. Separation of variability and uncertainty in QMRA models (so-called second-order models) has up to now rarely been made, a reflection of the fact that this can be a daunting task. However, neglecting the difference between them may lead to improper risk estimates (Nauta 2000) and/or incomplete understanding of the results (Vose 2000). Apart from these considerations, modeling variability has priority over modeling uncertainty. Furthermore, one must realize that in addition to the parameter uncertainty referred to here, uncertainty also includes scenario uncertainty and model uncertainty.

The second aspect is the considerable data need (*Risk assessment of food borne bacterial pathogens: quantitative methodology relevant for human exposure assessment* 2003). Data will be needed on environmental conditions (e.g., temperature, pH) and (handling) practices (e.g., duration of transport, storage) at the various processing steps. To validate the model, data are needed on N, unit size and P at the beginning and end of all steps. Another category of data that is needed concerns the description of the food pathway. Experience showed that a description in quantitative terms (number of animals and their destination or their origin (national, import), number and weight of carcasses and their destination or their origin, etc.) is not easily obtained. Moreover, when the model is to be used also to gain insight into risk reduction scenarios, data on alternative food pathways and/or steps will be needed. A third category of data is consumption data. For exposure assessment, data on prevalence and level of pathogens are not sufficient. Exposure can only be estimated if data on amount and frequency of food intake in the given population or subpopulation are available. It will be clear that in principle a large amount of data is needed to perform a risk assessment. Usually a large amount of data is available for any particular subject. However, these data were usually not collected for the purpose of risk assessment and therefore often of little use.

Hazard characterization

Exposure assessment is followed by hazard characterization, which can be implemented by an effect model. An effect model consists of a dose–response model, which translates the ingested dose into a probability of infection, and a disease-burden model, which estimates the probability of each of the relevant diseases, given infection. Each disease contributes to the disease burden, which can be expressed in an integrated measure, e.g. DALYs (see below).

The *E. coli* study mentioned above (Nauta et al. 2001) included not only exposure assessment, but also effect modeling, distinguishing three age classes in the human population (1-4 years, 5-14 years and 15+). The number of STEC O157 infections by steak tartare consumption per year in the Netherlands was predicted with the baseline model at 2,300, and the number of cases of gastroenteritis at about 1,300. The latter equals an incidence rate of 8 per 100,000 person years. This result can be compared with an independent point estimate of the total incidence of STEC O157-associated gastroenteritis in the Netherlands based on epidemiological data: 2,000 cases or 13 per 100,000 person-years. This would imply that a large fraction of the cases is a consequence of steak-tartare consumption. As many more routes of exposure to STEC O157 are known, the large attributable fraction of steak-tartare consumption (that is the high contribution to the total incidence), seems to be an overestimation. However, one should realize that comparing the model and epidemiological estimates is questionable, as the uncertainty in both these estimates is large.

Analysis of alternative scenarios shows that the uncertainty in prevalence and concentration of STEC O157 at farm level may have a large effect on the final model estimates. The same holds for uncertainty about growth and inactivation of STEC O157 on the carcass. In contrast, the effect of growth of STEC O157 during retail and domestic storage is negligible and the effect of advocating the consumption of 'well done' steak tartare patties is questionable. This suggests that intervention at farm level or at slaughter is more likely to be effective as a strategy to reduce STEC O157-associated risks than intervention at the consumer level.

We will use research on *Campylobacter* (Teunis and Havelaar 2000; Havelaar 2002) to illustrate effect modeling in some more detail. As stated above, one part of effect modeling is dose response modeling. Many models are available for dose response modeling, but data are scarce. A theoretically well defensible model is the hypergeometric model. The mathematical formula for this model can be derived on the basis of the following assumptions:

- each individual micro-organism is able to infect a human (the single-hit hypothesis);
- individual micro-organisms do no interact (hypothesis of independent action);
- the probability of infection per micro-organism varies (following a beta distribution), e.g. because of variation in pathogen virulence or host susceptibility.

Figure 4 shows the result of fitting the hypergeometric model to experimental data with human volunteers.

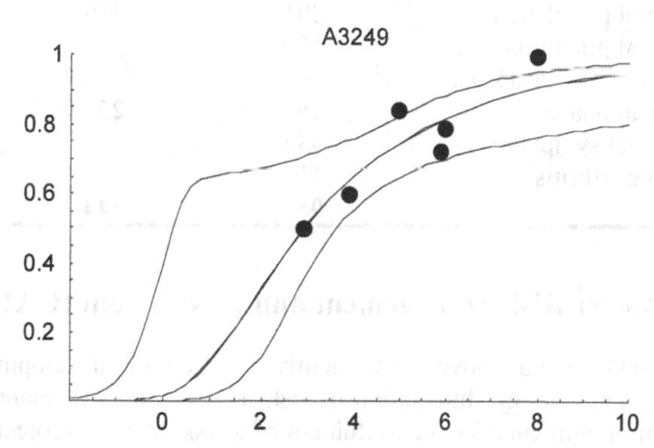

Figure 4. Hypergeometric dose–response model, fitted to data from Black et al. (1988) on *Campylobacter jejuni* A3249. X axis: ^{10}log of the dose; Y axis: fraction of infected persons. Middle curve: best fitting curve; outer curves: limits of the 95% confidence interval.

The second part of effect modeling is disease modeling: estimating the probability of diseases occurring once infection has taken place (Havelaar 2002). *Campylobacter* infection can remain without indications of disease, but frequently it will lead to gastroenteritis, characterized by diarrhea, stomach pain, fever and less frequently vomiting and blood in the stool. Less frequently, more serious disease symptoms will occur, of which the Guillain-Barré syndrome (GBS) and reactive arthritis (ReA) are the most important (Havelaar et al. 2000). Death can occur as a consequence of gastroenteritis (especially in the elderly) and as a consequence of GBS; no deaths have been reported as a consequence of ReA. Combination of the models for the probability of infection and the probability of disease in infected persons leads to a model for the probability of diseases as a function of the ingested dose.

An integrated measure for comparison of reductions in health status is the DALY (Disability-Adjusted Life Year) (Murray and Lopez 1996; Van der Maas and Kramers 1998). The principle is that mortality as well as morbidity is expressed in (healthy) life years lost: Years of Life Lost (YLL) and Years Lived with Disability (YLD), respectively. Thus life years can be lost by premature death (the loss is equal to the theoretical remaining life expectancy at the time of death, had the disease not occurred) and some proportion of time lived because disease or infection reduces the quality of life. This proportion depends on the severity of the disease.

An example of application of DALYs is an epidemiological study on the disease burden of thermophilic *Campylobacter* species in the Netherlands (Havelaar et al. 2000). The results are given in Table 1. It appears that the main determinants are acute gastroenteritis in the general population, gastro-enteritis related mortality and residual symptoms of Guillain-Barré syndrome.

Table 1. Disease burden due to infection with thermophilic *Campylobacter* spp. in the Netherlands (Havelaar et al. 2000).

Population	**YLD**	**YLL**	**DALY**
Gastroenteritis			
General population	291	419	710
General practitioner	159		159
Guillain-Barré syndrome			
Clinical phase	16	25	41
Residual symptoms	334		334
Reactive arthritis	159		159
Total	**959**	**444**	**1403**

Campylobacter Risk Management and Assessment (CARMA)

The studies cited above were mainly focused on development of the risk-assessment methodology. Integration of risk assessment, risk management and risk communication will enhance the usefulness of a risk-analysis project. In the CARMA project (Figure 5) (Havelaar 2002) this will be attempted. Risk assessment models will be integrated with economic models and policy analysis to provide an optimal basis for risk-management decisions. The social and health effects of disease will be expressed in the disease-burden model as DALYs, and in the economic model the costs due to disease will be calculated. In consultation with risk managers and stakeholders, intervention scenarios will be selected and it will be investigated in which way these will lead to changes in the risk-assessment model. Autonomic developments will be taken into account when choosing the intervention scenarios.

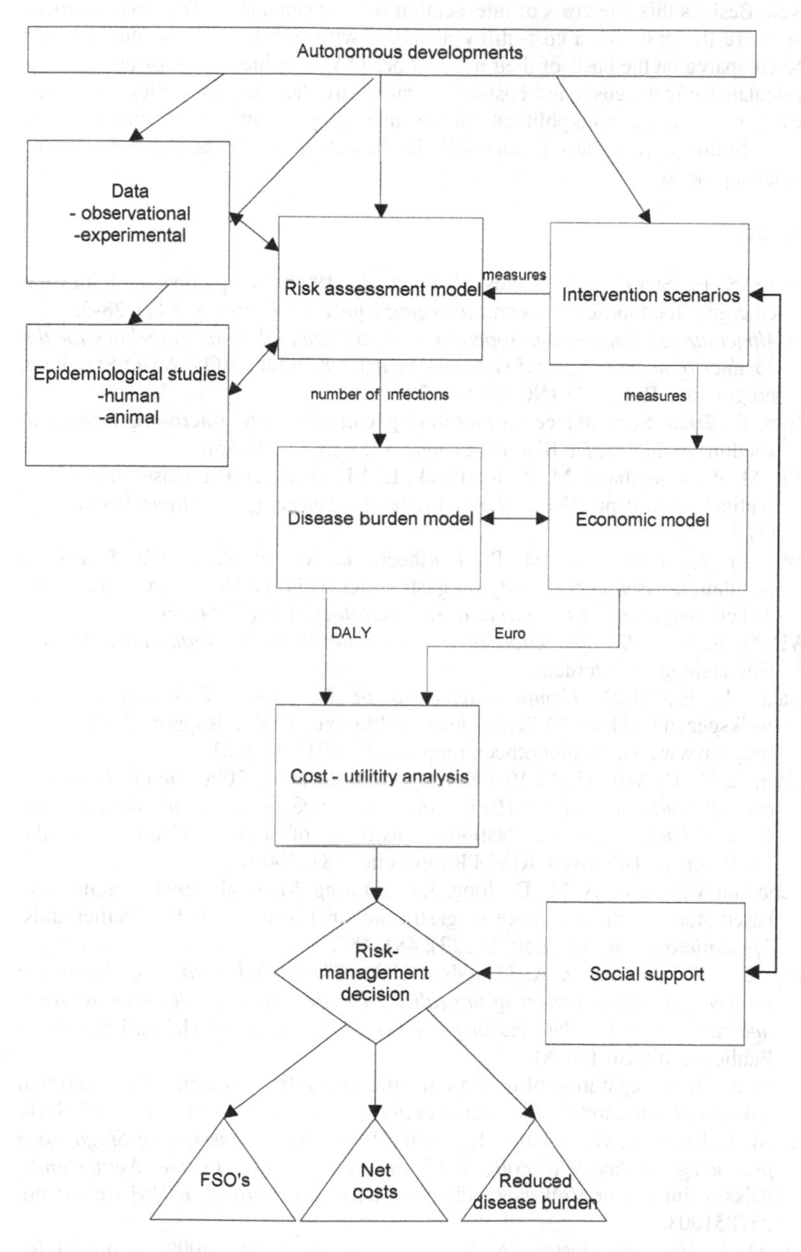

Figure 5. General setup of the CARMA (*Campylobacter* Risk Management and Assessment) project. FSO = Food Safety Objective.

Reduction of disease will lead to a reduction of disease burden and the costs involved. Besides this, the costs of intervention will be calculated. Together, all these estimates are the basis for a cost-utility analysis, with which different interventions will be compared on the basis of their net cost per DALY gained, the cost-utility ratio. The calculated effects, costs and cost-utility ratios are the basis for policy decisions. However, other social and political factors also play an important role in these decisions. In this project, these factors will also be listed, in a format that is useful for the decision process.

References

Altekruse, S. F., Stern, N. J., Fields, P. I., et al., 1999. Campylobacter jejuni--an emerging foodborne pathogen. *Emerging Infectious Diseases,* 5 (1), 28-35.

*Codex Alimentarius Commission. Appendix II: Draft principles and guidelines for the conduct of microbiological risk assesment*1998. Joint FAO / WHO Standards programme, Rome, ALINORM 99/13A.

De Boer, E., 2000. Surveillance en monitoring van pathogene micro-organismen in voedingsmiddelen. *De Ware(n)-chemicus,* 30 (3/4), 143-150.

De Wit, M. A., Koopmans, M. P., Kortbeek, L. M., et al., 2001a. Gastroenteritis in sentinel general practices, The Netherlands. *Emerging Infectious Diseases,* 7 (1), 82-91.

De Wit, M. A., Koopmans, M. P., Kortbeek, L. M., et al., 2001b. Sensor, a population-based cohort study on gastroenteritis in the Netherlands: incidence and etiology. *American Journal of Epidemiology,* 154 (7), 666-674.

De Wit, M. A. S., 2002. *Epidemiology of gastroenteritis in the Netherlands*. Ph. D., University of Amsterdam.

Havelaar, A. H., 2002. *Campylobacteriose in Nederland.* Rijksinstituut voor Volksgezondheid en Milieuhygiene, Bilthoven. RIVM Rapport 250911001. [http://www.rivm.nl/bibliotheek/rapporten/250911001.pdf]

Havelaar, A. H., De Wit, M. A., Van Koningsveld, R., et al., 2000. *Health burden in the Netherlands (1990-1995) due to infections with thermophilic Campylobacter species.* National Institute of Public Health and the Environment, Bilthoven. RIVM Rapport no. 284550004.

Hoogenboom-Verdegaal, A. M., De Jong, J. C., During, M., et al., 1994. Community-based study of the incidence of gastrointestinal diseases in The Netherlands. *Epidemiology and Infection,* 112 (3), 481-487.

Murray, C. J. L. and Lopez, A. D. (eds.), 1996. *The global burden of disease: a comprehensive assessment of mortality and disability from diseases, injuries, and risk factors in 1990 and projected to 2020. Volume I.* Harvard School of Public Health, Boston, MA.

Nauta, M. J., 2000. Separation of uncertainty and variability in quantitative microbial risk assessment models. *International Journal of Food Microbiology,* 57, 9-18.

Nauta, M. J., Evers, E. G., Takumi, K., et al., 2001. *Risk assessment of Shiga-toxin producing Escherichia coli O157 in steak tartare in the Netherlands.* Rijksinstituut voor Volksgezondheid en Milieu, Bilthoven. RIVM Report no. 257851003.

Nauta, M. J., Van De Giessen, A. W. and Henken, A. M., 2000. A model for evaluating intervention strategies to control salmonella in the poultry meat production chain. *Epidemiology and Infection,* 124 (3), 365-373.

*Proposed draft principles and guidelines for the conduct of microbiological risk assessment*2001. Joint FAO / WHO Food Standards Programme, Codex Committee on Food Hygiene, Rome, CX / FH 01/7.

*Risk assessment of food borne bacterial pathogens: quantitative methodology relevant for human exposure assessment*2003. Available: [http://europa.eu.int/comm/food/fs/sc/ssc/out308_en.pdf] (6 Mar 2003).

Teunis, P. F. and Havelaar, A. H., 2000. The Beta Poisson dose-response model is not a single-hit model. *Risk Analysis,* 20 (4), 513-520.

Van der Maas, P. J. and Kramers, P. G. (eds.), 1998. *Volksgezondheid toekomst verkenning 1997. III. Gezondheid en levensverwachting gewogen.* Rijksinstituut voor Volksgezondheid en Milieu, Bilthoven.

Van Pelt, W., De Wit, M. A. S., Van De Giessen, A. W., et al., 1999. Afname van infecties met Salmonella spp. bij de mens: demografische veranderingen en verschuivingen van serovars. *Infectieziekten Bulletin,* 10 (5), 98-101. [http://www.rivm.nl/infectieziektenbulletin/bul105/izboli3iz.html]

Vellinga, A. and Van Loock, F., 2002. The dioxin crisis as experiment to determine poultry-related campylobacter enteritis. *Emerging Infectious Diseases,* 8 (1), 19-22.

*Voedselinfecties*2000. Gezondheidsraad, Commissie Voedselinfecties, Den Haag. Publicatie / Gezondheidsraad no. 2000/09. [http://www.gr.nl/pdf.php?ID=164]

Vose, D. J., 2000. *Risk analysis : a quantitative guide*. 2 edn. John Wiley & Sons, Chichester, UK.

TRACEABILITY AND CERTIFICATION IN THE SUPPLY CHAIN

5

Technical and economic considerations about traceability and certification in livestock production chains

Miranda P.M. Meuwissen, Annet G.J. Velthuis, Henk Hogeveen and Ruud B.M. Huirne*

Introduction

Food-safety scandals such as the dioxin crisis in the poultry sector, the MPA crisis in the pork sector and the BSE crisis in the beef sector have created or strengthened consumers' belief that food can be unsafe. A major aspect of the scandals was that the contamination was not immediately detected. Furthermore, after detection the exact source of contamination was hard to find within a reasonable time. As a consequence, there was distrust in the safety of the food that was still in the food stores.

In January 2000, the European Commission outlined radical new principles for food safety in its White Paper on Food Safety (*White paper on food safety* 2000), a few months later specified in a proposal for new food-safety hygiene rules (*Health and consumer protection directorate-general, 2000*). These rules state, among others, that food safety is the primary responsibility of food producers. Linked to this, there is an obligation for non-primary food operators to implement HACCP (Hazard Analysis of Critical Control Points) systems and for farmers to implement sector-specific Codes of Good Hygienic Practice. Furthermore, it is stated that all food and food ingredients should be traceable and that proper recall procedures should be in place for food that might present a serious risk for consumers' health.

The food-safety hygiene rules do not mention the need for certification of 'good manufacturing practices'. Still, certification of the type of systems required by the hygiene rules is becoming increasingly important. Figure 1 illustrates the relationship between the food-safety hygiene rules issued by the European Commission and the (accredited) standards used for certification by certification services. The figure shows that (from left to right) the food-safety hygiene rules lead to regulatory standards at country level and, next, to food-safety and hygiene systems and traceability systems at company and chain level. National surveillance and control services monitor whether these systems fulfil the regulatory standards. Besides, the systems can be certified by public or private organisations using (accredited) standards.

This paper focuses on the right-hand side of Figure 1 and more specifically on traceability and certification. The goal of the paper is to analyse the status and perspectives of traceability systems and certification schemes and to review their potential costs and benefits. The following two sections describe purposes, requirements, status and perspectives of traceability systems and certification

* Institute for Risk Management in Agriculture, Wageningen University, Hollandseweg 1, 6706 KN Wageningen, The Netherlands

A.G. J. Velthuis et al. (eds), New Approaches to Food-Safety Economics, 41-54.

schemes, respectively. Then there is a section discussing potential costs and benefits, followed by a comprising section with the conclusions and an economic research agenda for the field of traceability and certification in livestock production chains.

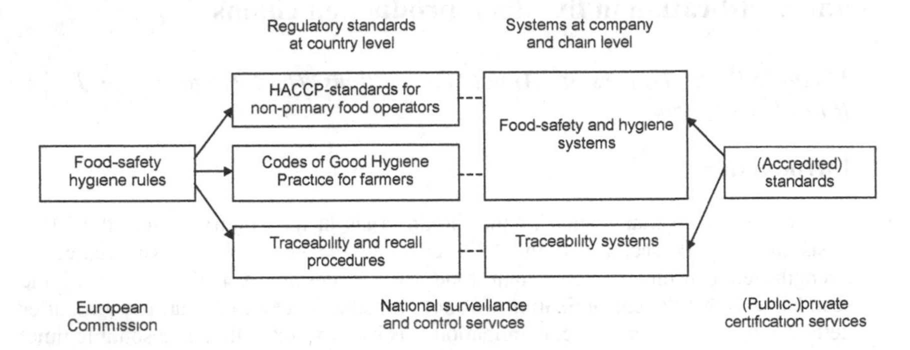

Figure 1: Relationship between the food-safety hygiene rules and (accredited) standards used for certification.

Traceability systems

Definition, purpose and requirements

A traceability system provides a set of data about the location of food and food ingredients along the production chain. Data relate to both the 'where' and 'when' issues. There are various relevant understandings. *Tracing* is the ability to trace food and food ingredients back along the production chain, i.e., from the end user to the producer and even to the suppliers of the producer. Tracing is aimed at finding the history of a product, for example to allocate the source of contamination. *Tracking* refers to the ability to track food and food ingredients forward along the production chain. Tracking can be used to find and recall products that might present a serious risk to consumers' health. *Identity preservation* is the set of measures taken to preserve and communicate the exact identity and source of food and food ingredients to the end user.

Traceability systems can be set up with different purposes in mind. For instance, to increase transparency in the production chain. More transparency is likely to increase consumers' trust in food safety due to the increased amount of information about, among others, production processes, food-safety controls, animals' living conditions and the use of medicines. Increasing transparency is also likely to enhance the actual level of food safety as a result of the improved information flows throughout the chain. Another purpose of implementing a traceability system can be to reduce the risk of liability claims: a proper traceability system is a valuable tool for companies to counterattack liability claims and to recoup claims from other participants in the production chain. Traceability systems can also be developed to improve recall efficiency. With an adequate system, the quality of recalls can be improved, which reduces costs and enhances the image of the production chain. These benefits can also be attributed to traceability systems that enhance the control of livestock epidemics.

For a traceability system to be adequate there are a number of requirements. First of all, all partners within the production chain should be identifiable – also small producers and hobby farmers. The latter is especially important if the traceability system is also used for the control of livestock epidemics (Disney et al. 2001). Secondly, there should be a unique animal identification system (McKean 2001), usually changed into an identification system for batches of animals as soon as the processing level is reached. Thirdly, an adequate traceability system requires a credible and complete (in the sense of what has been agreed on) information transfer along all participants of the production chain.

Current status

Three different types of traceability systems can be distinguished. These are outlined in Figure 2. In system "A", each link in the production chain gets its relevant information about the former link from the former link. The advantage of this type of system is that the amount of information to be communicated remains small, which reduces transaction costs. The disadvantage is that this system is largely based on trust. Each link has to trust the former link on the quantity and quality of the information passed. Furthermore, in case of an emergency, all links need a perfect administration in order to act fast.

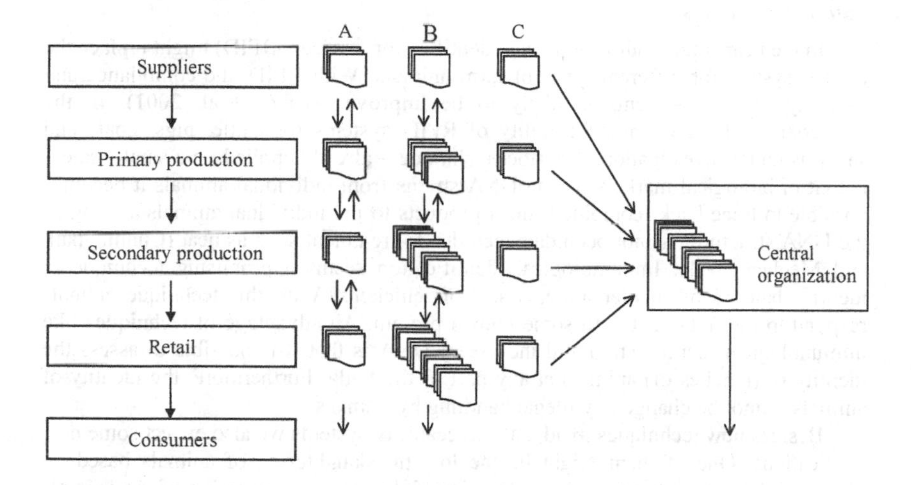

Figure 2: Traceability systems in production chains.

In system "B" each link gets the relevant information about all former links from the former link. With these systems, the speed at which tracking and tracing can be handled is much higher than with systems of type "A". Moreover, because each link in the chain receives all other information, the information can be controlled on completeness. Also the chain's transparency seems larger than with system "A". A disadvantage is that the amount of information to be transferred increases per link.

In the third type, system "C", each link of the production chain provides the relevant information to a separate organisation, which combines the information of all links in the whole production chain. Such organisation can solve the matter of trust. Also, tracking and tracing can in principle be carried out rapidly. Moreover, since the organisation is dedicated to the system, the danger that the system is not well maintained because of lack of time or other resources is minimised. On the other hand, total costs may be larger.

An example of a traceability system of type "C" can be found in the European beef industry. Due to the BSE crisis, the beef industry put into place a basic version of a traceability system: each package of beef contains information about the country of origin of the animal, the country of growing, the country of slaughtering and the country of butchering. To provide this information, individual countries have identification and registration (I&R) systems in place. In the Netherlands, for instance, each cow receives a unique life number at the moment of birth and two yellow ear tags on which this number is visible. The unique life number is registered at a central database together with some additional information, such as the unique farm code of the farm of birth. When the calf or cow leaves the farm, this has to be registered by both the delivering and the receiving party through the use of an automatic voice-dialling system. In this way, a cow can be traced back and tracked forward at any moment in time. In the UK, the ear tags are combined with a cow passport, which accompanies the cow for its whole life (Pettitt 2001). A same sort of system is also in place for pigs. However, pigs are registered on batch level, using earmarks with a unique farm number.

Future perspectives

In the near future radio-frequency identification devices (RFID) might replace the ear tag system for different types of farm animals. With RFID, the compliance and usability of I&R systems is likely to be improved (Ribó et al. 2001). In the Netherlands, the economic feasibility of RFID systems for cattle, pigs, goats and sheep is under investigation. A further technique – already applied on a small scale – is that of biological markers. Using DNA strains from individual animals it becomes possible to trace back (combined) meat products to the individual animals as long as the DNA structure has not been damaged due to treatment such as heat (Cunningham and Meghen 2001). Immunological identification seems a promising technique to identify batches of smaller animals such as chickens. With this technique animals respond to the treatment with some known protein. An advantage of techniques like immunological identification and the use of DNA is that it is possible to assess the identity of (batches of) animals at any part of the body. Furthermore, the identity of animals cannot be changed by illegal handling by humans.

Besides new techniques to advance traceability systems we also expect some new applications. One of them might be the logistic slaughtering of animals based on historical data about the prevalence of microbiological contamination of the animals or farms, for instance with respect to *Salmonella*. A further additional application includes more detailed assessments of animal-breeding values based on information about the production and offspring of individual animals. Traceability systems may also be used in the future for the inclusion of extra information, for example with respect to the primary production circumstances of animals. Such additional information enhances product differentiation and branding.

Certification

Definition, purpose and requirements

We define certification as follows: certification is the (voluntary) assessment and approval by an (accredited) party on an (accredited) standard. As this definition shows, certification is a very broadly used term. However, it certainly involves an assessment and an approval on some standard. The 'approval of good practice' distinguishes certification from the activities by national surveillance and control services (Figure 1), which do not go any further than only evaluating if implemented systems fulfil the regulatory standards.

Certification is, in general, voluntary. However, there are also cases in which it is 'quasi-voluntary'. For example, if it is a customers' requirement or if there are price disadvantages from not participating in a certification scheme (Payne et al. 1999; Bredahl et al. 2001). Also risk-financing organisations, such as banks and insurance companies, may require some form of certification in their underwriting policy (Bullens, Van Asseldonk and Meuwissen 2002; Skees, Botts and Zeuli 2002). In relation to the certifying party and the standard used for certification, it can be stated that if an accredited standard is used, the certification procedure needs to be carried out by an accredited party (Tanner 2000). All other type of standards can be certified by either accredited parties, (other) third parties, such as product boards and interest groups, or customers (also called 'second parties'). Figure 3 gives an overview of the various certifying and certifiable parties in the livestock production chain. The dotted line linking "accredited party" with "other third party" and "second party" refers to the fact that an accredited party can be employed by any other party to carry out certification audits.

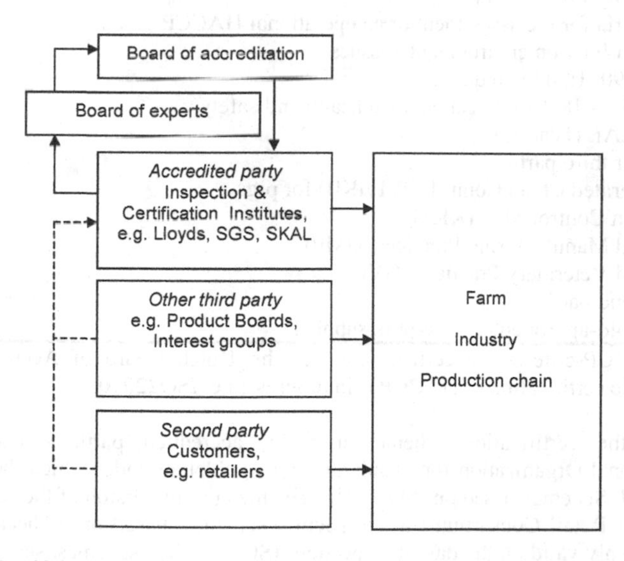

Figure 3: Certifying parties (left) and certifiable parties (right) in livestock production chains.

The purpose of certification is to reach a defined performance and to make this perceptive to stakeholders. Stakeholders may include consumers, other customers, governments, risk-financing parties such as banks and insurance companies, and society as a whole. Also the company itself can be a stakeholder, since certification of food-safety and traceability systems gives organisations a tangible approval of good practice and a tool for due-diligence defence in case of product safety (see for instance (Buzby and Frenzen 1999; Henson and Holt 2000)).

For stakeholders to regard certification as a valuable tool, they must trust the certification scheme as well as the certifying party. Also, there should be regular tests or audits (usually specified in the certification scheme) to verify whether the certified party still reaches the agreed performance level.

Current status

Many certification schemes fit in the context of the food-safety hygiene rules (Figure 1). For each of the certifying parties, Table 1 lists a number of examples.

Table 1: Examples of certification schemes from different certifying parties and, if relevant, underlying ISO guidelines.

Certification schemes by	**ISO guide**
Accredited parties	
BRC (British Retail Consortium)	39 (65?)
SQF-1000 (Safe Quality Food on farm level)	65
SQF-2000 (Safe Quality Food on industry level)	65
EKO (organic)	65
EUREP-GAP (Good Agricultural Practices)	65
Sector-specific Codes of Good Hygiene Practice	65
Identity Preservation	65
Criteria for the assessment of an operational HACCP system*	62
ISO 14001 on environmental issues	62
ISO 9001:2000 on quality	62
OHSAS 18001 on occupational health and safety	62
HALAL (Islamic)	62
Other third parties	
Integrated Chain Control (PVE/IKB) for pigs	-
Chain Control Milk (KKM)	-
Good Manufacturing Practices (GMP)	-
Good Veterinary Practices (GVP)	-
Second parties	
"Ahold-approved organic-pork supplier"	-

*The HACCP-criteria are certifiable under the Dutch Board of Accreditation. A world-wide certification of HACCP is in progress, i.e. ISO22000.

For the certification schemes used by accredited parties, also the ISO (International Organization for Standardization) guideline under which they resort is mentioned. Schemes based on ISO 39 (such as the current version of the scheme from the British Retail Consortium) are inspection schemes, based on a checklist and in principal only valid on the day of inspection. ISO 65-based schemes, such as EKO (a Dutch certificate for organic products), are product-certification schemes in which products as well as processes are tested on specified standards. Certification schemes based on ISO 62, such as ISO 9001:2000, are system-certification schemes. They use system requirements to evaluate complete management systems. The Dutch Integrated Chain Control (PVE/IKB) and the Chain Control Milk (KKM) are examples of certificates issued by 'other third parties'. An example of a customer-issued certification scheme is the 'Ahold-approved organic-pork supplier'. Most schemes have some requirements with regard to traceability. A scheme specifically focusing on the issue of traceability is Identity Preservation.

Since product- and system-certification schemes do not use straightforward checklists, individual auditors' interpretations become increasingly important with these types of schemes. For instance, in the EKO certification scheme for livestock production it is stated that 'pig-breeding systems should allow sows direct access to the soil (..) except where bad weather or unsuitable soil conditions make housing preferable'. An auditor has to judge whether the housing circumstances at a farm fulfil these requirements. A clearer requirement in the EKO certification scheme is that 'the maximum number of laying hens that can be kept in one group is 3000'. As an example of system certification, the HACCP criteria require 'to identify hazards' and subsequently 'to carry out risk analyses'. Controls by the Board of Accreditation (Figure 3) prevent large interpretation differences between individual auditors.

In Figure 3 it can be noted that also production chains can be certified. However, if a production chain consists of multiple legal entities, as is often the case in livestock production chains, there are only limited opportunities for certification. From a second-party point of view there is a practical limitation of certifying production chains: if a chain consists of multiple entities there is no single addressing point. For 'other third parties' there are in principle no limitations (compare KKM and PVE/IKB), but these certificates are not based on accredited standards. Attaining accredited certification schemes seems to be increasingly important for production chains for reasons of credibility. However, accredited certification schemes based on ISO 62, such as the HACCP criteria and ISO9001:2000, only apply to legal entities. Furthermore, accredited certification for each individual chain participant (either under ISO 62- or ISO 65-based certification schemes) is very costly for production chains in which there are many small enterprises (see for instance (Unnevehr and Jensen 1999; Taylor 2001)). As an alternative, product-market organisations can be set up and certified by an accredited party. This is illustrated in Figure 4.

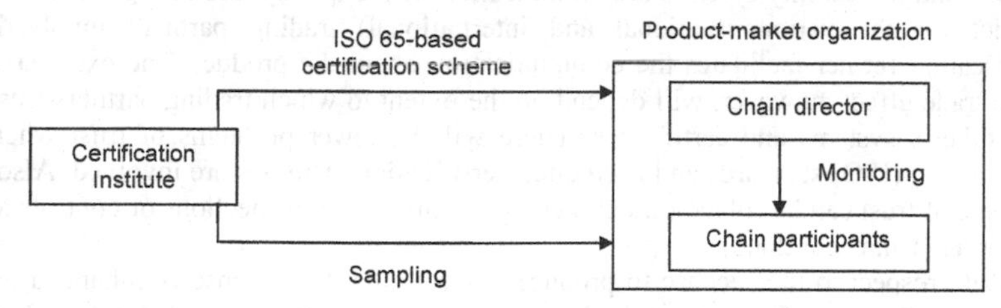

Figure 4: Accredited certification of a product-market organization.

Accredited certification of a product-market organisation has a number of characteristics. Firstly, there is one chain director, for instance a slaughterhouse, product board, or, even possible, a single farmer. Secondly, the certification scheme is issued to the chain director for the full scope of the chain. Thirdly, chain participants are monitored by the chain director and sampled by the certification institute. And fourthly, certificates are restricted to ISO 65-based schemes since these do not require single legal entities. A rather similar approach to certify farms was described by Mazé, Galan and Papy (Mazé, Galan and Papy 2002) under the term "pyramidal certification systems".

Future perspectives

Elaborating on current developments, we argue that the importance of certification will further increase. Governments increasingly shift responsibilities to companies under the initial assumption that certified products, processes and systems are in conformity with regulatory standards. In practice, standards used for certification even go beyond the legislative provisions. Certification will also become more important when due diligence becomes increasingly important.

We furthermore expect that the role of retail organisations will increase, leading to standards such as EUREP-GAP (Good Agricultural Practices), introduced by the European retailers' organisation. Benchmark models such as GFSI (the Global Food-Safety Initiative, set up by the International Food Business Forum) assist in 'standardising the standards' so that it remains possible to oversee the increasing number of standards and certification schemes.

Costs and benefits

Table 2 gives an overview of the costs and benefits of food-safety and hygiene systems (FS&H), traceability systems (T&T) and the certification (Cert.) of these systems, from both the industry and consumer perspective. Food-safety and hygiene systems are included here separately in order not to mix the costs and benefits of a system on the one hand and the certification of such a system on the other hand. The table does not distinguish between the various participants of the production chain, or between the various types of systems and certification schemes.

Since most items in Table 2 are fairly straightforward or were already discussed in the previous sections, we focus here on the three aspects: the positive effect on trade, the enhanced license to produce and the price premium. These aspects are mentioned important for food-safety and hygiene systems, as well as traceability systems and certification. The magnitude of the discussed aspects is debatable.

The positive effect on trade is attributed to the fact that food-safety and hygiene systems and traceability systems are an indication of the quality and background of a product for the various (national and international) trading partners involved. Certification further facilitates the communication about the product. The exact size of the trade effect, however, will depend on the extent to which trading partners trust each other's systems and certificates. There will be fewer problems of trust when international (ISO) standards and accredited certification institutes are involved. Also, the issue of trust can be solved (at extra costs) by in-country inspections or controls at the border (Unnevehr 2000).

With respect to the 'license to produce', stating that this license is enhanced by introducing the type of systems and schemes under consideration is probably true. But the question is for how long: as soon as the public is used to the upgraded market, new requirements are likely to be introduced. Discussions on the 'license to produce' became actual in countries like Germany and the Netherlands, following the epidemics of BSE, foot-and-mouth disease and classical swine fever.

Table 2: Potential costs and benefits of food-safety and hygiene systems (FS&H), traceability systems (T&T), and certification (Cert.), subdivided into industry and consumers*.

	Costs	Benefits
Industry		
FS&H	– Implementation: development, training, capital purchases – Maintenance: verification and validation, analyses, record keeping, operating processes	– Improved internal efficiency: improved agreements, explicitness about tasks, responsibilities and authorities of employees – Less failure, i.e. recall, closure, scrap and liability costs – Positive effect on trade – Enhanced 'license to produce' – Price premium
T&T	– Implementation: transforming production process, less flexibility, automation, extra storage capacity, production materials, personnel and documentation – Maintenance: audits	– Increased transparency of production chain – Reduced risk of liability claims – More effective recalls – More effective logistics – Enhanced control of livestock epidemics – Positive effect on trade – Enhanced 'license to produce' – Price premium
Cert.	– Implementation – Maintenance: audits	– Lower transaction costs from supplier Identification, contract negotiation, verification and enforcement – Enhanced access to insurance and finance – Effectuated due diligence – Positive effect on trade – Enhanced 'license to produce' – Price premium
Consumers		
FS&H	– Price premium	– Enhanced level of food safety
T&T	– Price premium	– Enhanced level of food safety
Cert.	– Price premium	– Enhanced level of food safety – Lower transaction costs

*Based on, among others: (Caswell and Hooker 1996; Roberts, Buzby and Ollinger 1996); (Bredahl and Holleran 1997); (Crutchfield et al. 1997); (Early and Shepherd 1997; Jensen, Unnevehr and Gomez 1998); (Jensen and Unnevehr 1999); (Golan et al. 2000); (Henson and Holt 2000); (Unnevehr 2000); (Bredahl et al. 2001); (Bullens, Van Asseldonk and Meuwissen 2002).

In relation to consumers paying a price premium for food-safety related systems and certification schemes, there is also uncertainty involved. In general, food safety is perceived as important, in particular in developed countries (Unnevehr 2000). Even more, consumers are generally willing to pay an extra price for safer food (see for instance (Henson 1996)). However, literature with respect to consumers' interest in the underlying systems resulting in the safer food, becomes less convincing. Walley, Parsons and Bland (Walley, Parsons and Bland 1999) state that it cannot be concluded that consumers are willing to pay for quality assurance. Gellynck and Verbeke (Gellynck and Verbeke 2001) found that traceability is perceived as important, but especially with respect to functional attributes, such as the monitoring of chains and individual chain-participants' responsibility in case of abuses. With regard to certification, Vastola (Vastola 1997) concluded that "consumers' attitude towards certification is twofold: while declaring their willingness to pay a higher price for a certified produce, when faced with the choice among different produces it is economic convenience that matters, not the presence of certification". Blend and Van Ravenswaay (Blend and Van Ravenswaay 1999) also support this conclusion. Literature thus supports our impression that it is not clear-cut whether participants in the livestock production chain receive a price premium for implementing food safety, hygiene and traceability systems and for certifying them.

Besides the debatable magnitude of some specific aspects, a relevant consideration in estimating the size of all costs and benefits listed in Table 2 is the definition of the reference point, or, the 'without project alternative' (Belli et al. 2001). A main aspect in this relates to the type of systems already in place. For instance, additional costs and benefits of implementing a HACCP system can expected to be less if there is already some sector hygiene code in place. Also other characteristics of the livestock production chain, such as the structure of the chain under consideration, will affect the size of costs and benefits (Golan et al. 2000). For instance, in the Netherlands implementing a traceability system in an integrated chain such as the veal chain will be less costly than implementing a similar system in a patchy and dispersed chain such as the dairy-cattle sector. A further relevant aspect is the size of farms and industries involved. Costs are likely to be non-linear, i.e. there is a possible comparative disadvantage for small and medium-size companies, as also denoted by Unnevehr and Jensen (Unnevehr and Jensen 1999) and Taylor (Taylor 2001).

Conclusions and economic research agenda

From our findings we conclude that there are numerous perspectives for traceability systems in livestock production chains and that the importance of certification schemes in these chains is likely to increase further. We furthermore conclude that there is in general much more attention for the more technical issues of traceability and certification than for economic considerations. We therefore would like to recommend that future developments be guided more by economic analyses than by technical prospects. In this respect we propose the following research agenda. The agenda (presented in arbitrary order) covers multiple economic disciplines:

Economic design of traceability systems

Research about traceability systems generally focuses on technical aspects. Additional research is needed to include also important economic aspects. For instance, what is the desired level of detail of a traceability system, i.e., is it efficient to be focused on a system 'as detailed as possible' or is there some break-even point? Relevant questions include whether it is necessary to be able to trace back to individual animals or is tracing back to the herd level also sufficient? Furthermore, is it necessary to track forward to all individual customers who received specific products, or would it also be adequate to work with day and batch codes so that products can be recalled at a higher level? A further consideration relates to the acceptable level of risk of traceability systems. 'Acceptable level of risk' is a common term in food-safety systems, but apparently not in traceability systems. A relevant question in this respect is whether production chains need to be able to track and recall all products in one way or the other, or whether they could for instance rely on some alert system at the end of the chain.

Distribution of costs and benefits of traceability along the production chain

Transaction costs of traceability systems are likely to increase as one moves along the production chain (Bredahl and Holleran 1997). However, also benefits may increase, for example as a result of increased selling opportunities. A better insight into the distribution of the costs and benefits along the production chain makes it possible to allocate price premiums (if any) accordingly. The relevance to do so was already mentioned by Verbeke (Verbeke 2001).

Optimization of incentives for participating in traceability systems

Although a proper allocation of price premiums along the production chain may be an incentive for most participants of the production chain to participate in traceability systems, this may trigger not every chain participant. This may, for example, be the case with farmers. They face possibly high claims from larger companies further in the chain while they have only limited financial means themselves to counterattack such claims. Incentive problems may especially arise when there is a risk of cross-contamination during processing. Solutions may be found in accredited certification schemes at the farm level, for instance through product-market organisations (as illustrated in Figure 4). Such schemes enhance farmers' opportunities to prove due diligence. A group of farmers that deserve special attention in terms of incentives to participate in traceability systems are hobby farmers. Hobby farmers can considerably influence the introduction and spread of livestock epidemics and the speed at which epidemics can be controlled. They are, however, not participating in certification programs and they are probably not affected by economic incentives such as a lower price for their products if traceability requirements are not fulfilled.

Reconsideration of liability and recall-insurance schemes

The increasing number of product-liability claims and the fact that they are 'moving backwards in the chain' require a reconsideration of liability- and recall-insurance schemes for all participants in the production chain. On the one hand, the need for insurance coverage is reduced because of the increased implementation of the type of systems and certification schemes discussed in this paper. On the other hand, adequate insurance coverage seems to be an increasingly important prerequisite for the long-term continuation of individual farms and companies because of the – small probability but high consequence – risk related to liability claims and recalls. Risk analyses supporting insurance studies are enhanced through the increased amount of traceability and food-safety information that is available.

Communication about food-safety-related systems and certification with consumers

The problem that consumers' willingness to pay for safer food and the implementation of food-safety systems and certification schemes is not straightforward, may be a problem of communication. Key research questions are:

- What information should be presented on the label (e.g., 'HACCP' or 'guaranteed safe', 'from Umbria' or 'fulfills our national standards')?
- Are consumers able to distinguish between various labels?
- Are there alternative ways of communication (see also Frewer (Frewer 2000))?
- What is the interaction with other consumer concerns, such as environmental aspects and animal welfare?
- What market segments can be distinguished? More insight into such questions would improve the communication with consumers about food-safety issues.

References

Belli, P., Anderson, J. R., Barnum, H. N., et al., 2001. *Economic analysis of investment operations: analytical tools and practical applications*. The World Bank, Washington, DC.

Blend, J. R. and Van Ravenswaay, E. O., 1999. Measuring consumer demand for ecolabeled apples. *American Journal of Agricultural Economics,* 81 (5), 1072-1077.

Bredahl, M. E. and Holleran, E., 1997. Food safety, transaction costs and institutional innovation. *In:* Schiefer, G. and Helbig, R. eds. *Proceedings of the 49th Seminar of the European Association of Agricultural Economists, 19-21 February 1997, Bonn, Germany*. European Association of Agricultural Economists, Bonn, 51-67.

Bredahl, M. E., Northen, J. R., Boecker, A., et al., 2001. Consumer demand sparks the growth of quality assurance schemes in the European food sector. *In:* Regmi, A. ed. *Changing structures of global food consumption and trade*. Economic Research Service-USDA, Washington, DC, 90-102. Agricultural and Trade Report no. WRS-01-1.

Bullens, A. C. J., Van Asseldonk, M. A. P. M. and Meuwissen, M. P. M., 2002. *Risk management in agriculture from a mutual insurance perspective. 13th Conference on International Farm Management in Agriculture, 7-12 July 2002, Wageningen, The Netherlands*.

Buzby, J. C. and Frenzen, P. D., 1999. Food safety and product liability. *Food Policy,* 24 (6), 637-651.

Caswell, J. A. and Hooker, N. H., 1996. HACCP as an international trade standard. *American Journal of Agricultural Economics,* 78 (3), 775-779.

Crutchfield, S. R., Buzby, J. C., Roberts, T., et al., 1997. *An economic assessment of food safety regulations: the new approach to meat and poultry inspection.* Economic Research Service-USDA, Washington, DC. Agricultural Economic Report no. 755.

Cunningham, E. P. and Meghen, C. M., 2001. Biological identification systems: genetic markers. *Revue Scientifique et Technique,* 20 (2), 491-499.

Disney, W. T., Green, J. W., Forsythe, K. W., et al., 2001. Benefit-cost analysis of animal identification for disease prevention and control. *Revue Scientifique et Technique,* 20 (2), 385-405.

Early, R. and Shepherd, D., 1997. A holistic approach to quality with safety in the food chain. *In:* Schiefer, G. and Helbig, R. eds. *Proceedings of the 49th*

Seminar of the European Association of Agricultural Economists, 19-21 February 1997, Bonn, Germany. European Association of Agricultural Economists, Bonn, 391-400.

Frewer, L., 2000. Risk perception and risk communication about food safety issues. *Nutrition Bulletin,* 25, 31-33.

Gellynck, X. and Verbeke, W., 2001. Consumer perception of traceability in the meat chain. *Agrarwirtschaft,* 50 (6), 368-374.

Golan, E. H., Vogel, S. J., Frenzen, P. D., et al., 2000. *Tracing the costs and benefits of improvements in food safety: the case of hazard analysis and critical control point program for meat and poultry*. Economic Research Service-USDA, Washington, DC. Agricultural Economic Report no. 791.

Henson, S., 1996. Consumer willingness to pay for reductions in the risk of food poisoning in the UK. *Journal of Agricultural Economics,* 47 (3), 403-420.

Henson, S. and Holt, G., 2000. Exploring incentives for the adoption of food safety controls: HACCP implementation in the U.K. dairy sector. *Review of Agricultural Economics,* 22 (2), 407-420.

Jensen, H. H. and Unnevehr, L. J., 1999. HACCP in pork processing: costs and benefits. *In:* Unnevehr, L. J. ed. *Economics of HACCP : new studies of costs and benefits. Proceedings of a NE-165*. Eagan Press, St Paul, MN, 29-44.

Jensen, H. H., Unnevehr, L. J. and Gomez, M. I., 1998. Costs of improving food safety in the meat sector. *Journal of Agricultural and Applied Economics,* 30 (1), 83-94.

Mazé, A., Galan, M. and Papy, F., 2002. The governance of quality and environmental management systems in agriculture: research issues and new challenges. *In:* Hagedorn, K. ed. *Environmental cooperation and institutional change: theories and policies for European agriculture*. Edward Elgar, Cheltenham, UK, 162-182.

McKean, J. D., 2001. The importance of traceability for public health and consumer protection. *Revue Scientifique et Technique,* 20 (2), 363-371.

Payne, M., Bruhn, C. M., Reed, B., et al., 1999. Our industry today: on-farm quality assurance programs; a survey of producer and industry leader opinions. *Journal of Dairy Science,* 82 (10), 2224-2230.

Pettitt, R. G., 2001. Traceability in the food animal industry and supermarket chains. *Revue Scientifique et Technique,* 20 (2), 584-597.

Ribó, O., Korn, C., Meloni, U., et al., 2001. IDEA: a large-scale project on electronic identification of livestock. *Revue Scientifique et Technique,* 20 (2), 426-436.

Roberts, T., Buzby, J. C. and Ollinger, M., 1996. Using benefit and cost information to evaluate a food safety regulation: HACCP for meat and poultry. *American Journal of Agricultural Economics,* 78 (5), 1297-1304.

Skees, J. R., Botts, A. and Zeuli, K., 2002. The potential for recall insurance to improve food safety. *International Food and Agribusiness Management Review* (special issue), 99-111.

Tanner, B., 2000. Independent assessment by third-party certification bodies. *Food Control,* 11 (5), 415-417.

Taylor, E., 2001. HACCP in small companies: benefit or burden. *Food Control,* 12 (4), 217-222.

Unnevehr, L. J., 2000. Food safety issues and fresh food product exports from LDCs. *Agricultural Economics,* 23 (3), 231-240.

Unnevehr, L. J. and Jensen, H. H., 1999. The economic implications of using HACCP as a food safety regulatory standard. *Food Policy,* 24 (6), 625-635.

Vastola, A., 1997. Perceived quality and certification. *In:* Schiefer, G. and Helbig, R. eds. *Proceedings of the 49th Seminar of the European Association of Agricultural Economists, 19-21 February 1997, Bonn, Germany*. European Association of Agricultural Economists, Bonn, 281-304.

Verbeke, W., 2001. The emerging role of traceability and information in demand-oriented livestock production. *Outlook on Agriculture,* 30 (4), 249-255.

Walley, K., Parsons, S. and Bland, M., 1999. Quality assurance and the consumer: a conjoint study. *British Food Journal,* 101 (2), 148-161.

White paper on food safety, 2000. Available: [http://europa.eu.int/comm/dgs/health_consumer/library/pub/pub06_en.pdf] (6 Mar 2003).

6

Traceability and certification in food quality production – a critical view

*Gerhard Schiefer**

Introduction

Traceability and certification in quality-assurance systems are issues which are widely discussed and which are the basis for major initiatives in policy, agricultural interest groups and agri-food industries.

This paper outlines some arguments on these issues, which are meant to initiate a discussion on the appropriateness of principal developments and on the needs for engagements of research either to support or to change their course of direction.

It is the opinion of the author that present discussions and initiatives have lost focus on the key visions and messages of quality management as it developed during the last twenty years. It largely disregards developments in research and experiences from industries, and builds on the knowledge status of many years ago. This opinion is based on ongoing developments with sector-oriented quality-assurance systems in various countries like, e.g., the German initiative Q+S (Budde and Richard 2002).

Difficulties in discussions are partly due to differences in focus. The focus of quality-management research is primarily on enterprises and on enterprise-level management approaches. If a sector is involved in quality discussions, it usually builds on clearly separated supply chains and brands like, e.g., in the automobile industry. The view is closely linked to the enterprise approach and covered by traditional quality-management research.

The agri-food sector is different in many aspects, which makes it difficult to link traditional enterprise-oriented quality-management approaches with quality-assurance requirements of the sector. Its infrastructure is characterized by a large number of rather small production enterprises (farms) on one end and internationally operating industries and retail chains for input delivery and food-product sales on the other end. The dependency on natural production environments sets limits to the control of production environments and makes the delivery of a comprehensive food program or the use of food-product ingredients dependent on a diversity of production sources on a global scale. Basic food products are similar in appearance irrespective of their source and quality, which reduces consumers' ability to make informed decisions.

However, difficulties in the utilization of enterprise-oriented quality-management approaches for the solution of sector quality-assurance problems should not result in the negligence of quality-management research results, but challenge quality-management research to find appropriate ways for the transfer of its accumulated knowledge from an enterprise into a sector view.

---

* *Department of Agricultural Economics, University of Bonn, Meckenheimer Allee 174, D-53115 Bonn, Germany*

A.G. J. Velthuis et al. (eds), New Approaches to Food-Safety Economics, 55-60.

It is the purpose of this paper to analyse some of the problems in sector-oriented quality-assurance approaches and to identify options for the transfer of established quality-management concepts and experience into the sector. The discussion has to some extent to rely on logical arguments and expert conclusions from general observations of developments, as scientific empirical studies on these subjects are still rare.

The next chapter focuses on the difference between enterprise-oriented quality-management and sector requirements. A key critical success factor for sector initiatives is 'consumer trust' and its support through various means like, e.g., certification. These issues are discussed in following chapters. The last but one chapter integrates the arguments into a basic framework for sector quality development in which traceability and certification have their place, but with less prominence than suggested by the current agenda of public discussions.

Enterprise vision and sector initiatives: the conflict

The term "quality management" describes a dynamic management perspective with an offensive and strategic vision. It evolved from half a century of initiatives initiated by Demin (1986) and continued by authors like Juran (1988), Crosby (1979), Ishikawa (1982), Taguchi (1986) and others (for a historical overview see Brocka et al. (1992)). Its core messages focus on development paths towards

- meeting customer expectations (described as quality),
- motivating people to engage in self-control and quality improvement, and
- improving process reliability and process efficiency.

Current discussions and initiatives on tracking, traceability and quality certification throughout agriculture and the agri-food sector are primarily defensive. Their focus is on external inspection and the establishment of sector-wide monitoring infrastructures, which are supposed to guarantee certain baseline quality characteristics and, in case of failures in delivery, to identify possible sources and causes for further action. However, the establishment of monitoring and inspection systems free of risk is an illusion. Any failure to deliver quality in any part of such a system – which is supposed to deliver quality in food and perceived by consumers as risk-free or with a low risk – might discredit the whole sectoral monitoring infrastructure, while consumers will be unaware of any other accomplishments. Previous experiences with universal monitoring systems which had to be abandoned after food-quality failures support this view. A case in point is the German sector initiative 'Kontrollierte Produktion' (controlled production), which was abandoned after the BSE crises reached the country.

Professional evaluations of control systems usually focus on probabilities of failures derived from expert judgement or statistical experience and on probability-based definitions of food safety. However, customer evaluations are based on customers' subjective judgements of risk, which develop from different perspectives. It is sometimes argued that 'quality' and 'food safety' are commodities attached to products which could be handled like any other physical commodity. This argument builds on the assumption that one could link the commodities to undisputed objective values. While this may be feasible in certain environments, in Europe it is certainly not.

The main tasks in the marketing of any quality management system are to

- reduce a gap between customers' risk perception and experts' risk judgement and

– reach a level of risk perception where 'distrust' or unacceptable risk switches to 'trust' or acceptable risk.

The key issue: trust

The principal approaches for building trust are the experiences and 'beliefs' of the consumers. Experience takes time to develop and can easily be shattered by singular events of system failures, whatever their statistical probability, if they are perceived as 'severe' by consumers. The feature of traceability does not improve or protect trust acquired through experience, i.e., the negative effects of system failures cannot be counterbalanced by the existence of a traceability feature.

Trust built on beliefs could develop faster on arguments convincing to the target group. And this type of trust remains more indifferent to singular system failures if the beliefs do not only refer to the system itself and its monitoring and control structure but also to underlying supporting elements, which may not have failed, i.e., where the foundation of system trust remained unshattered. In this context a traceability feature could have a positive effect if it demonstrated the stability of the supporting elements. This would allow a delineation of system failures from the supporting elements.

It is obvious that a combination of experience and beliefs would be the most stable basis for sustainable trust and the target to be approached. However, market pressures towards food-quality and food-safety guarantees force the sector to implement concepts which, in a first step, build on beliefs and lay the ground for the subsequent development of experience. A starting point for beliefs could be the 'appropriate' communication of objective probabilities for system failures derived from expert judgement or statistical experience. It is sometimes argued that a low failure probability could be communicated as 'safe food'. This might be the appropriate communication approach if consumers translate it back to 'low failure probability'. However, if consumers translate it back to 'zero failure probability', any system failure will discredit the communication system. A communication system in the sensitive area of trust needs to build on a thorough analysis of consumers' perceptions of communication concepts. It is our opinion that in food-safety discussions, probabilities from expert judgement or statistical experience still receive too much attention as compared to consumers' perception of risk.

It is common understanding that principal supportive elements are (a) trust in the appropriateness of processes and (b) trust in people. The first element has been the essence of quality management from the very beginning. In this approach, trust builds on impressions of customers that a certain quality-assurance system is based on a progressive system approach where quality guarantees are combined with continuous and reliable quality assurance and improvement efforts. Such impressions can be developed for integrated food supply chains with clearly distinguishable brands or for clearly identifiable sub-sectors like the sub-sector for organic food.

Trust in the appropriateness of processes may receive additional support through reverse backward tracing. In this approach, backward tracing does not refer to failures but is used for the provision of supporting quality information like, e.g., the display of animals' living conditions or the display of controls to consumers. This type of personal monitoring of guarantees by consumers may serve as a substitute for process guarantees provided by, for instance, certification.

The second principal supportive element, trust in people, builds on human relationships and longstanding established experience with the trustworthiness of people who provide personal guarantees on the effectiveness of controls or the reliability of processes. Examples, which were well publicized during the BSE crises, are

- trust in the safety of organic food which did not build on traceability but on images of quality and reliability based on trust in the dedication of people, and
- common references to local butchers with their direct and longstanding supplier–customer relationship with consumers.

Certification as a Means for Trust

It is a common approach to use third parties to support promises of guarantees, which build on control systems or appropriate process-organization and process-improvement systems through auditing and certification procedures. In food marketing, the value of such procedures depends on their ability to generate trust. Certification might support the development of experience or beliefs if critical customers understand a certification approach and is itself accepted as a trustworthy approach.

However, it is doubtful that certification in a general sector (network) control system can fulfil all the expectations. Network systems build on generally accepted quality levels which, as a consequence, tend to result in low-level guarantees. Supply chains in network systems evolve from actual market operations with changing partners and usually do not involve clearly visible 'trusted' supplier-customer relationships. In this scenario, quality-improvement initiatives face co-ordination problems and are of rather limited value to participants and customers. Participants have difficulties to disconnect from the network in case of system failures anywhere in the network. Customers have to build their trust on a system, where guarantees primarily have to depend on the inspection system but less on widely accepted personal responsibilities or process organizations. This increases the risk of failures and reduces the value of guarantees for customers.

Certification in this application scenario lacks improvement potential and the personalization element, which might shield it against the loss of trust in case of failures. Its main application value might be in the establishment of a new system as long as no failures occur.

Sustainable and effective certification must allow clearly identifiable segmentation through, e.g., branding of products from clearly specified supply chains. Branding based on clearly identifiable participants supports self-control, motivation and competitive quality improvement. Closed supply chains are the natural basis for high-quality branding, high value of certification, and high differentiation potential in case of system failures within the network as a whole.

However, while a general closed-chain approach might be appealing from a quality-assurance point of view, it is not a feasible solution for the agri-food sector as a whole. The dependency of agricultural production on natural production environments leads to fluctuations in quantity and quality, and, in turn, to conflicts between the needs of markets for continuous delivery of a certain quality and the actual service possibility. This requires sector buffers and a sector organization, which is best modeled by a sector network with chains or enterprises as member units.

A basic framework for sector quality development

The various arguments in the preceding chapters can be grouped into a framework for sector-oriented quality-assurance systems in the agri-food sector. It views the sector as a network of interconnected enterprises for the production and delivery of food products. The framework involves the following main features:

- Establishment of a hierarchical control and certification system which allows a clear, understandable and accepted identification of different levels of food quality and food safety.
- Visible delineation of co-operating sub-networks for the introduction of stricter quality claims and for improved utilization of quality-supporting elements. This would facilitate a dissociation of sub-networks from the general network in the perception of customers in case of failures of the general control system.
- Utilization of quality-supporting and trust-generating elements by clearly identifiable sub-networks: (a) Personalization may be introduced through the organization of identifiable sub-networks consumers can identify themselves with (e.g., regions), the implementation of reverse backward tracing, or the activation and communication of dedication by people or groups involved. The latter requires, to be effective, a motivating incorporation of enterprises from all stages of the production and delivery process. (b) The activation and communication of processes with a convincing built-in continuous quality-improvement feature.

The principal framework needs to be translated into sector activities. The identification of sub-networks through branding is not enough. The framework asks for a different understanding of branding as a comprehensive ('total') approach for quality assurance. It also changes our view on the development of market organizations towards a further electronic integration. It is best modeled by a network of interconnected but separable trade platforms which link participants of sub-networks (Lazzarine, Chaddad and Cook 2001; Hausen, Helbig and Schiefer., 2001). Similar approaches need to be designed for other aspects of sector developments, which could together form a comprehensive quality-assurance model for the agri-food sector.

Conclusion

Developments in food markets ask for sector quality-assurance systems. However, the focus of traditional quality-management research is on enterprises and not on sectors. The transfer of quality-management principles into sector environments and the analysis of customer reactions on failures in food safety provide the basis for a sector quality-assurance model. While the principal elements of such a model seem to be clear, the translation into activities and organizational infrastructures still needs to be done.

References

Brocka, B. and Brocka, M. S., 1992. *Quality management: implementing the best ideas of the masters*. Irwin, Homewood.

Budde, F. J. and Richard, A., 2002. Qualität und Sicherheit. *Landwirtschaftliches Wochenblatt für Westfalen und Lippe* (Special issue).

Crosby, P. B., 1979. *Quality is Free: : the art of making quality certain.* McGraw-Hill, New York, NY.

Deming, W. E., 1986. *Out of the crises.* Massachusetts Institute of Technology, Cambridge, MA.

Hausen, T., Helbig, R. and Schiefer, G., 2001. Networked trade platform. *In:* Schiefer, G., Helbig, R. and Rickert, U. eds. *E-Commerce and Electronic Markets in Agribusiness and Supply Chains. Proceedings of the 75th Seminar of the EAAE, February 14-16, 2001, Bonn, Germany.* Universität Bonn, ILB-Verlag, Bonn, 213-222.

Ishikawa, K., 1982. *Guide to quality control.* Asian Productivity Organization, Tokyo.

Juran, J. M., 1988. *Juran on planning for quality.* The Free Press, New York.

Lazzarine, S. G., Chaddad, F. R. and Cook, M. L., 2001. Integrating supply chain and network analysis: the study of netchains. *Journal of Chain and Network Science,* 1 (1), 7-22.

Taguchi, G., 1986. *Introduction to quality engineering.* Asian Productivity Organization, Tokyo.

FARM-TO-TABLE RISK ANALYSIS AND HACCP

7

Food-system risk analysis and HACCP

Helen H. Jensen[*]

Introduction

Controlling food-safety risks has received increased public and private attention in the last several years. The increased attention stems from several sources. New scientific evidence has linked food-borne hazards to human health; consumers now demand higher levels of food safety because of higher income; and consumers have increased preferences for "food-safety attributes" in foods due to greater awareness of food-borne problems. Other changes in food markets and supply contribute to the rise in food-safety problems. The increase in consumers' food spending to away-from-home sources of food has shifted control of food preparation and service away from consumers' control. Recurring food recalls, well-publicized food-borne health risks and threats (including new concerns about bio-terrorism) and large-scale, centralized production and distribution channels have heightened consumers' awareness of the vulnerability of their food supply to food-borne hazards. In addition, as technical and trade barriers to food trade have fallen, trade in food products has increased and imported foods represent a growing source of foods in many countries. This trade also introduces new sources of risk in the food supply through easier transfer of food-borne hazards and plant and animal diseases.

The changes in food markets and increased awareness of food-borne illness as a public-health concern have led to increased public and private demand for food safety. Private certification (both self and third party), related contracting schemes, and quality-control systems have become important methods of quality assurance in food marketing and trade. In the public sector, there has been increased interest in policies and regulation to assure public health and, at the same time, make efficient use of public resources. Significant innovation in both the private and public sector has led to the development of methods and institutions to improve safety of products and assure consumers of the safety of the food supply.

Food-safety assurance is at the heart of the food-safety problem. Private markets often fail to provide adequate food safety because information costs are high, detection often very difficult, and the nature of the contamination is complex. Underlying many of the food-safety failures is the existence of externalities, or costs not borne by those whose actions create them. Externalities tend to arise when strong dependencies govern relationships between economic agents, and when the production environment is not sufficiently well understood to allow market-based solutions (Hennessy, Roosen and Jensen 2002). Strong dependencies between agent decisions exist in food supply chains. Microbial agents are widespread, can lead to significant hazards, are often difficult to detect, and can re-enter the food supply chain, even after control at earlier stages. When firms are not able to capture fully the

[*] *Department of Economics, Iowa State University, 578 Heady Hall, Ames, IA 50011, USA*

A.G. J. Velthuis et al. (eds), New Approaches to Food-Safety Economics, 63-77.

returns from incorporating costly controls of product hazards, they lack the incentive to implement production methods to assure a safer product.

Many governments have taken a new approach to ensure the safety of the food supply that rests on an increased "scientific basis" to food-safety control and risk-assessment framework. This emphasis can be illustrated by control of microbial hazards in products of animal origin, and the focus in this paper is on meats, poultry and animal products. Mandated use of Hazard Analysis of Critical Control Points (HACCP) systems has placed focus on verifiable control in the food production process. In the United States, HACCP was mandated for seafood in 1994 (Procedures for the safe and sanitary processing and importing of fish and fishery products: final rule, docket no. 93N-0195 1995), for meat and poultry in 1996 (Pathogen reduction hazard analysis and critical control point (HACCP) systems: final rule, docket no. 93-016F 1996), and for fresh fruit juice in 2001 (Hazard analysis and critical control point (HACCP): procedures for the safe and sanitary processing and improving of juice: final rule 2001); and regulations taking a risk-based approach have been mandated for shell-egg handling (Food labeling, safe handling statements, labeling of shell eggs; refrigeration of shell eggs held for retail distribution: final rule 2000).

The European Union directive 93/43, effective in December 1995, requires member states to adopt a HACCP approach in obliging food companies to follow HACCP principles in their production process (Grijspaardt-Vink 1995; Ziggers 1999). The companies, themselves, are responsible for monitoring their food safety, although final authority lies with the national authorities (Bunte 1999). The recent European Communities (EC) White Paper on Food Safety (*White paper on food safety* 2000) identifies the guiding principles for food-safety policy to include:

- taking a comprehensive, integrated approach throughout the food chain;
- identifying responsibility for food safety through the food chain – from farm to table;
- basing food-safety policy on the foundation of risk analysis in the design of standards; preventing hazards through the use of HACCP;
- and implementing traceability to assure monitoring as required to protect the safety of materials and inputs.

The growing use of HACCP as a sanitary standard in international trade led the Codex Alimentarius to adopt guidelines for HACCP in 1993, and to incorporate HACCP into food-hygiene codes starting in 1995 (Whitehead and Orriss 1995), cited in (Unnevehr and Jensen 1999)).

Despite the general acceptance of HACCP by regulators and international agencies, there are several issues concerning its mandatory imposition and use as a regulatory tool, especially in light of greater emphasis in taking a comprehensive and integrated approach throughout the food chain. First there is concern about how effectively the use of HACCP will control or eliminate some food-safety hazards when applied in food-safety regulations aimed at the firm (Hathaway 1995). Second is the question of whether the nature of HACCP as a regulatory tool changes when moving from a firm to a farm-to-table approach, when dependencies govern relationships among agents. And third is the question of how cost–benefit analysis can be integrated into farm-to-table risk analysis. This is to allow for evaluation of alternative interventions or systems of prevention and to investigate whether a systems-based approach is likely to lead to lower costs of food-safety control than one based on controls in each stage of the process.

This paper begins with an overview of the types of costs and benefits used in regulatory cost–benefit analysis. The focus is primarily on the approaches that have been used to measure costs. I draw from Unnevehr and Jensen (Unnevehr and Jensen 2001) and summarize what is known from studies of the costs of control of microbial hazards. This is followed by a discussion of the systemic nature of risk in the food system. By exploring the nature of systemic risk in the provision of food it is possible to gain a better understanding of how moving to evaluation of risks in an *integrated* framework can guide alternative food-safety control policies (Hennessy, Roosen and Jensen 2002). The alternative types of risk in the food system are used to motivate consideration of linkages in farm-to-table risk analysis and approaches to control the risks. Two examples illustrate the farm-to-table approach and use of HACCP in control systems. The final section uses the examples and taxonomy of system risk to consider the nature of cost evaluation and implications for costs in systems of control.

Approaches to measuring social costs and benefits

The Environmental Protection Agency (EPA) recently published guidelines for cost–benefit analysis of environmental regulation (*Guidelines for preparing economic analysis* 2000). These provide a useful starting point. As described there, and discussed elsewhere in this proceedings, the valuation of benefits assigns monetary values to the market and non-market benefits of regulation. The social benefits of improved food safety include reductions in risks of morbidity and mortality associated with consuming contaminated food. There are several accepted ways of assigning valuations, including welfare measures of willingness to pay/willingness to accept. In the food-safety area, quantification of the physical effects of the regulation or policy analysis on reduced morbidity or mortality can lead to assignment of values through the use of cost-of-illness methods. These methods estimate the change in explicit market costs associated with reduced incidence of an illness (or death) due to a policy or regulation. The uncertainties in the assignment of benefits may be significant due to lack of data.

The costs that result from regulation include four types of costs: real-resource compliance costs, government regulatory costs, social-welfare losses, and transitional social costs, as shown in Table 1 (*Guidelines for preparing economic analysis* 2000; Unnevehr and Jensen 2001). The costs incurred by firms, which must change production processes in some way to meet new standards or regulations, are termed real-resource compliance costs. Costs can be either fixed costs that require an investment over several years or variable costs that are incurred with each unit produced. Costs can be very concrete and easy to measure, such as the purchase of new equipment like the steam pasteurizer used in beef-packing plants, or more difficult to measure, such as changes in labor organization to monitor temperatures. The simplest kind of cost analysis is an accounting for these costs within a static framework (e.g., so many plants pay so much extra per unit of output). Government regulatory costs include the governmental costs to administer, monitor and enforce the food-safety policies. This may include the costs of inspectors, plant monitoring and testing.

The direct costs to firms lead to other changes in markets, such as social-welfare losses from higher consumer prices for meat products or increased costs of litigation, or transitional social costs, such as possible firm closings due to inability to meet standards competitively (Just, Hueth and Schmitz 1982). In measuring the latter two categories, both the distribution of real-resource costs and the adjustments to these costs are taken into account more fully. Adjustments may lead to lower costs over time as firms find more efficient ways to comply with standards, and understanding such adjustments is important for comparing regulatory alternatives. Furthermore, the distribution of costs both between consumers and producers and among different kinds of producers and consumers will have important implications for public policy.

Table 1: Examples of Social-Cost Categories

Social-cost category	**General examples**	**Food-safety examples**
Real-resource compliance costs	– Capital costs of new equipment – Operation and maintenance of new equipment – Change in production processes or inputs – Maintenance changes in existing equipment – Changes in input quality, such as skilled labor – Changes in costs due to product quality, can be positive or negative	– Steam pasteurizer – Additional water needed for rinses – More frequent cleaning – – Training of employees in HACCP procedures – Lower quality of product with reduced pesticide use
Government regulatory costs	– Federal, state or local government costs to administer, monitor and enforce new policies	– Inspectors – Mandated testing – Regulatory reporting costs
Social-welfare losses	– Higher consumer and producer prices leading to changes in consumer and producer surplus	– Higher prices for meat products – Higher insurance costs against recalls
Transitional social costs	– Legal/ administrative costs – Firm closings – Unemployment – Resource shifts to other markets – Transactions costs – Disrupted production	– Regional shifts in production – Small meat-processing plants shut down – Reduced stock value due to recalls

Adapted from Exhibit 8-2, in U.S. EPA "Guidelines for Preparing Economic Analysis" (2000). Based on Unnevehr and Jensen (2001).

Measuring direct compliance costs and their partial equilibrium impact on the market in question is usually the focus of regulatory analysis. Economists have extended this analysis in some cases to look more generally at impacts on several markets or at general equilibrium impacts in both factor and output markets. For example, Unnevehr, Gomez and Garcia (Unnevehr, Gomez and Garcia 1998) examined how HACCP costs would affect the three major meat-product markets differently, due to differences in the incidence of costs and resulting substitutions in demand among beef, pork, and chicken. These substitutions reduced the total welfare cost of the regulation. Another example is the general equilibrium analysis of HACCP by Golan et al. (Golan et al. 2000), who found that costs of implementation were almost fully passed through to households as a reduction in income (more than offset by a reduction in health-care costs on the benefit side). The distribution of costs and benefits varied among household types, with the greatest net benefits going to households with children.

These kinds of modeling efforts are useful for illuminating the long-run effects of the regulation and its resulting costs. Hayes et al. (Hayes et al. 2001) show that, although the effects of a ban on antimicrobial growth promotants in US swine feed would increase costs to producers by US$ 6 per pig head initially; the costs would fall over the 10-year period examined and profits would recover some. This is due to increased output prices with smaller supplies. Such dynamics are important in determining incentives for innovation and compliance.

Findings related to costs of regulating microbial hazards

Microbial hazards are naturally occurring organisms. They can enter food products throughout the food-supply/production chain, and once present, they can grow in numbers. Therefore control at one level does not assure control at subsequent levels; and lack of control at one level has consequences for the following stages in the food chain. This makes hazard control and the design of regulation more complex; it also complicates economic analysis of the costs of control. Unnevehr and Jensen (Unnevehr and Jensen 2001) review recent findings with respect to costs of control and HACCP. This section draws on their review. HACCP systems substitute process control (that includes significant data collection, monitoring and management in the production process) for the costs of testing end product. The HACCP controls are motivated by emphasis on prevention and the use of easily accessible indicators in efforts to reduce food-safety hazards (Unnevehr and Jensen 1996; MacDonald and Crutchfield 1996)

As applied in the US to the seafood, meat/poultry and juice industries; and in the EU to food processors and to the feed industry, specific HACCP plans are not mandated, but are incorporated in a regulatory framework that shifts responsibility for control of hazards to the firm level. Individual firms can develop plans relevant to their particular product mix and plant situation. The flexibility in this type of regulation means that it is difficult to estimate its costs *ex ante*. For example, it is unclear what kind of changes in production processes might result from HACCP implementation.

Because *ex ante* costs are difficult to estimate and controversial in the food industry, there has been considerable interest in estimating HACCP costs as the regulations are implemented. A number of studies have been undertaken of HACCP (see the collection of studies in (Unnevehr 1999), and it is now possible to make some *ex post* comparisons and generalizations, although more definitive answers will only emerge after longer experience. Studies of the costs of pathogen reduction show that both FSIS and FDA underestimated the costs of HACCP in their *ex ante* analyses. For example, Jensen and Unnevehr (Jensen and Unnevehr 1999) estimate that modifications of pork slaughter processes to reduce pathogens would cost US$ 0.20 to 0.47 per carcass, substantially more than the FSIS estimate of US$ 0.0056 for process modifications (Crutchfield et al. 1997). Antle (Antle 2000) analysed past costs of quality improvement in the meat industry. He extrapolated that a 20% improvement in safety would have additional costs in the range of 1 to 9 US$ cents per pound of product, which is several times larger than the FSIS estimates of less than one one-hundredth of a cent per pound.

Recent studies show that the marginal costs of pathogen reduction are increasing and suggest that complete control is quite costly. Jensen, Unnevehr and Gomez (Jensen, Unnevehr and Gomez 1998) found that pathogen-control marginal-cost curves are steeply increasing in both beef and pork. Costs rise from US$ 0.20 to 1.40 per beef carcass and from 3 to 25 US$ cents per pork carcass as pathogen reduction increases from one log to 4 logs.[i] Narrod et al. (Narrod et al. 1999) find rising costs of *E. coli* control in beef-packing plants – costs rise from 5 to 45 US$ cents per carcass as contamination is eliminated from 30% to 100% of production. Both of these studies emphasize that there is a frontier of efficient control technologies and technology combinations that provides least-cost pathogen reduction.

To date, actual costs incurred by meat and poultry firms likely are small relative to total costs and product prices. They may be around 1 to 2% of current processing costs (Jensen and Unnevehr 1999). And, although costs are small on average, they may still be enough to shift the distribution or scale of production at the margin. Small firms' costs are likely to rise proportionally more than large firms' with the implementation of HACCP due to large up-front investments in developing and implementing a HACCP plan. Furthermore, large firms frequently have more in-house resources at their disposal for design and implementation (e.g., meat scientists on staff; diagnostic labs) and therefore have lower transactions costs in implementing a HACCP plan.

However, the major difficulty in assigning costs to regulation is that firms face a mix of market and regulatory incentives in adopting food-safety measures. Certain markets increasingly demand evidence of hazard control from their suppliers and this provides motivation beyond the minimum prescribed by regulation. Martin and Anderson (Martin and Anderson 1999) report widespread adoption of HACCP and/or food-safety control procedures among US food-processing firms. Almost 70% of large plants have a HACCP plan for at least one product; a majority of these firms also carry out food-safety procedures associated with HACCP, such as monitoring temperatures of raw ingredients. Bunte (Bunte 1999) finds that in the Dutch food sector, HACCP tends to be implemented in the more concentrated industries, characterized by economies of scale. If market incentives drive firms to adopt food-safety practices then it is not clear to what extent additional food safety is a result of regulation or how to assign costs to the this component.

The adoption of tighter food-safety controls at one part of the food chain is likely to create incentives that are passed back to suppliers through the marketplace. The experience in Europe indicates that food processors and retailers are increasingly looking for assurances of food safety from their suppliers, creating incentives for improved safety throughout the food chain. In the United Kingdom, the passage of "due diligence" laws has forced food retailers to ask their suppliers for certification of hazard management (Henson and Northen 1998). In the US, such contracts tend to be motivated entirely by market incentives and there is less reported evidence that regulation has played a role.

The review of costs associated with microbial-hazard control shows inherent flexibility in the adaptation of HACCP control systems to the regulatory use. At the firm level, the evidence shows that the marginal cost of food-safety improvements is likely to be rising. Also, there are both private as well as regulatory incentives for improving food safety at the firm level. The question now is what are the cost implications of taking a farm-to-table systems-based approach to food-safety risks in the food chain?

The food system

Food-safety failures often stem from problems that are systemic in nature. The systemic failures occur in production systems characterized by interconnected stages in production and inputs, and this interconnectivity gives rise to the technological potential for failures. At the same time incentive problems provide the economic potential for failures (Hennessy, Roosen and Miranowski 2001; Narrod et al. 1999). The mixing of meat from a number of farm sources at the packer, processing or intermediary levels illustrates both, the interconnectivity in inputs and stages of production, and incentive problems. Ground meat may come from many different animal/farm sources. Problems that occur from the farm, or in handling of a single animal, can easily spread through the food product in the plant. Furthermore, when intermediaries co-mingle beef from several sources, failure in one large batch can quickly spread to consumers in a large geographic area (Hennessy, Roosen and Jensen 2002). Testing of products at different stages is often difficult (and rapid tests are not available). Incentive problems occur because it is difficult for packers to reward farmers for care taking, and farmers have no incentive to take additional care in production or transport to reduce the likelihood of problems at the packer level. Nor do packers that sell product to intermediaries that co-mingle beef from several sources have market incentives to adopt technologies that reduce pathogens in the plant source.

Figure 1 illustrates the type of systemic risks that can occur due to the interconnectivity of the food system. The figure illustrates the nature of failure in the system in the case when the cause of a failure is known, and when it is not known. Suppose that there are three retailers (or restaurants) that source from two providers. In Figure 1 the three retailer nodes are on the right. Arrows indicate the direction of product flow. Retailer *r1* sources from provider *p1* only, retailer *r3* sources from provider *p2* only, while retailer *r2* sources from both *p1* and *p2*. The circular arrows at the retailer nodes indicate that the retailers also provide some of their own inputs.

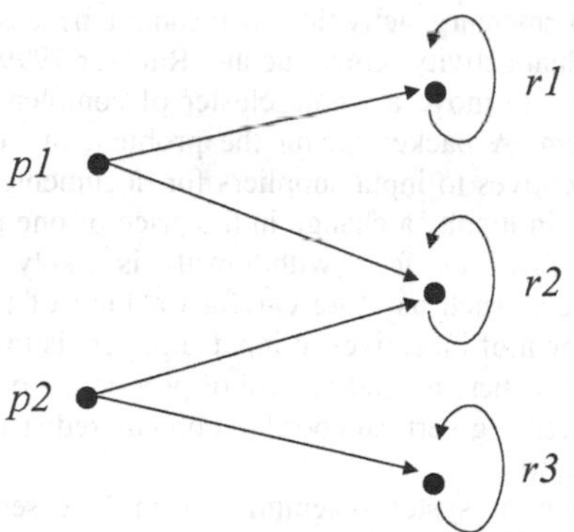

Figure 1. Node diagram of retailers and providers (from Hennessy, Roosen and Jensen2002)

If an illness arises at *r1*, then, without any information on the source of the problem, nodes *r1*, *r2*, and *p1* will have to close for quality audits. In contrast, if problems were detected at node *r2*, then the whole system would have to close down for audit. Node *r2* is the node most strongly connected among all the nodes in the system, and the systemic risk associated with a problem that becomes evident there is most severe. In comparison, if the cause of the problem were known, the system losses might be smaller (and would never be larger). If the problem was observed in *r2*, but also the cause was known to be *r2*, the losses would be limited to this single node.

A variant on this case is when the cause is known but mixing occurs. Figure 2 shows the system where there is contamination in an ingredient used in producing products *b1* and *b2*. If contamination developed in a meat or feed source, it would spread throughout the system even though the cause is known. Products would need to be removed from the system from all sources that received the ingredient due to mixing.

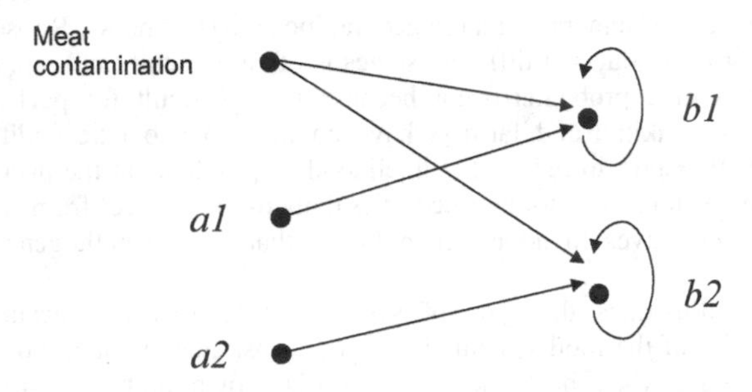

Figure 2. Node diagram of ingredient mixing (adapted from Hennessy, Roosen and Jensen 2002)

Interconnectivity may also give rise to complementarities in input use (care in one area may increase the likelihood of taking care in other aspects of production). The presence of complementarities among activities means that there may be benefits that arise from complementary activities that cannot be assigned to the marginal product of any individual activity (Goodhue and Rausser 1999). A change in the cost of one activity is likely to move a whole cluster of complementary activities in the food production system. A packer facing the problem of downstream risks might choose to provide incentives to input suppliers for documented production practices. With complementarity in inputs, a change in the price of one practice (e.g., incentive paid by the packer firm for feed withdrawal) is likely to bring along other complementary practices, such as more careful tracking of transportation practices. An alternative to payment of incentives to input suppliers is to purchase control of the input supply (i.e., shift ownership and control of production or transport to the packer firm). In this case, increasing vertical coordination can redistribute the risks and rents associated with reduced risk.

The interconnectivity in systems stemming from the cases presented in Figures 1 and 2 can be represented by identification of connections and information about the source of contamination problems. Probabilistic Scenario Analysis (PSA) and closely related Fault Tree Analysis (FTA) are tools that have been used to account for multiple events and the probability of any event occurring in the food production system (Roberts, Ahl and McDowell 1995). The PSA makes use of information on links in the food chain and events that may compromise the safety of the food: the type of hazard, the different ways it enters the food chain (e.g., the specific link and linkages), and the full list of other, expected events. The "links in the food chain" are specialized, self-contained activities that are connected to events that determine the human health outcome. An "event tree" summarizes this information (see, for example (Roberts, Ahl and McDowell 1995)). The PSA or FTA approach takes account of various linkages in the food system at a point in time, probabilities of occurrence and all associated probabilities of failure (or, alternatively, effectiveness of control). The high-risk pathway (most likely) becomes a likely candidate for control analysis.

Today, the system-wide risk-assessment framework is being applied in several studies, including ones on the potential for BSE in the United States (Cohen et al. 2001), the risk of Shiga-toxin-producing *E.coli* O157:H7 in steak tartare (Nauta et al. 2001) and on the risk of occurrence of *E.coli* O157:H7 in ground beef (2001). To illustrate, the USDA *E.coli* O157:H7 study provides a comprehensive evaluation of risk of illness from *E. coli* O157:H7 in US ground beef based on available data and assignment of the risk due to contamination throughout the farm-to-table continuum. The distributions of risk, including information on the variability and uncertainty in the assignment, account for full information on the risks and control measures. The exposure assessment includes production, slaughter and preparation stages. Cattle shipped to slaughter may carry threshold levels of the pathogen. The probability that *E. coli* O157:H7 contamination will occur in cattle at slaughter depends on whether (and how likely it is that) cattle carry the pathogen at the production (farm) level and whether the pathogen is detected at entry to the slaughterhouse. The slaughter operation is the second stage in the food chain or processing system. Later stages in the system occur through food preparation (processing and fabrication, distribution and transport, wholesale/retailing and finally at the consumer level with cooking and consumption). In the food production system, each of these stages offers potential for contamination or recontamination.

In principle, information on the probabilities and paths in the production system can be used to assign expected costs to various control options, and identify the most cost-effective mitigation options. By identifying combinations of lowest-cost interventions to achieve various levels of improved safety, the analyst can articulate optimal strategies. This approach combines risk outcomes and economic cost criteria to identify dominant solutions (McDowell et al. 1995). The outcome and cost-dominance approach underlies the models used to evaluate beef processing (Jensen, Unnevehr and Gomez 1998; Narrod et al. 1999) and pork processing (Jensen and Unnevehr 1999) that identify the cost-efficient combinations of interventions. In principle, however, such prescriptive economics is more likely to depend on a combination of methods from decision theory, risk analysis and economics (McDowell et al. 1995). Demands for data to support such analyses are very large. Although the PSA/FTA and farm-to-table risk-assessment approaches describe system linkages in food production, they give limited guidance for identifying strategies to reduce hazards across the whole system. They fail to account for incentives that may lead to different behaviors and choices of technologies and controls among stages.

Two examples

The complexity of most food production today suggests the importance of considering food-safety problems from the systems perspective. Two examples illustrate the potential for farm-to-table risk analysis and related cost analysis. The first is the action plan developed by the US FDA, FSIS and Animal & Plant Health Inspection Service (APHIS) to eliminate *Salmonella enteritidis* (SE) illness due to eggs (*Egg safety from production to consumption: an action plan to reduce salmonella enteritidis illnesses due to eggs* 1999). Underlying the action plan was a risk-assessment model. The risk-assessment model indicated that multiple interventions would achieve more reductions in SE illness than would a single point of intervention. The use of a risk-assessment approach allowed combining information about the risk, sources of risk and potential for controls throughout the egg production system and identified potential sites for intervention. The identified advantage of multiple interventions suggested following a broadly based policy approach across stages of production, instead of focusing on a single stage of production.

Figure 3, from the President's Council on Food Safety (*Egg safety from production to consumption: an action plan to reduce salmonella enteritidis illnesses due to eggs* 1999), shows the stages of egg production and the agencies responsible at each stage. The action plan identifies a set of activities at each stage. Producers and packer/processors can choose between two strategies designed to give equivalent performance in terms of reduction in SE at the egg production and packer/processor stages. The first strategy (Strategy I) focuses efforts on farm-level testing and egg diversion; the second strategy (Strategy II) directs more resources to the packer/processor level and includes a lethal treatment, or "kill step" (and HACCP plan) at this stage. Both strategies include common features of regulatory presence on the farm (e.g., control of chicks from SE flocks) and at the packer/processor (e.g., mandated prerequisite programs of sanitary controls, washing). In addition to the interventions at production and packer/processor stages, the action plan sets refrigeration standards for the distribution and retail stages to ensure that reductions in SE are preserved at later stages in the food supply chain. The flexibility offered to the industry in choosing between strategies for control at the producer and packing/processor levels allows for development of incentive structures consistent with the overall objectives of eliminating SE illnesses. The action plan identifies explicitly performance measures (output standards) to be used (e.g., reduced illnesses, SE isolates and number of SE outbreaks) and the responsible agency for each stage in the farm-to-table continuum.

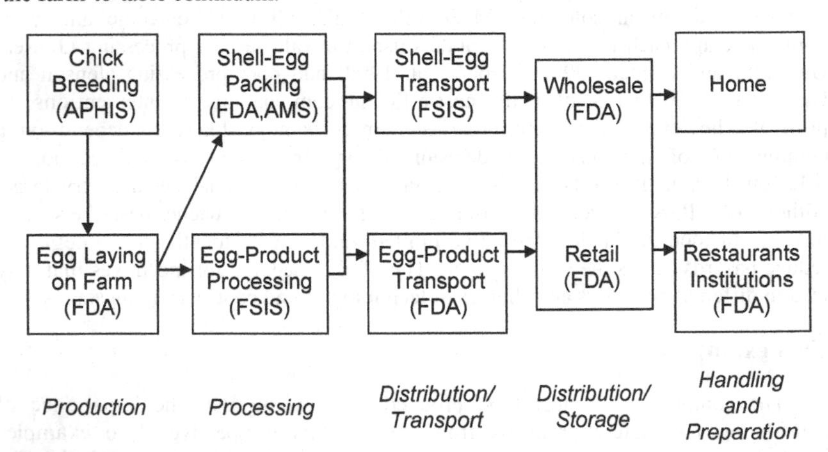

Figure 3. Egg Safety from Production to Consumption

The action plan for SE in eggs provides a good example of how a system-wide approach might be used. In this case, the systems approach facilitated the development and co-ordination of public and private strategies across the egg production system. The risk-assessment model focuses on the desired public-health outcome. The plan allows industry flexibility in developing and coordinating incentives across stages (production and processing/packing). Costs incurred under this systems approach are likely to be smaller than when interventions focus on only one point in the food chain. This is an example of how risk assessment can interface with economic incentives to achieve lower costs of controls through market incentives.

The second example is the recent study in Europe of *Salmonella in Pork* (Wong and Hald 2000). The study was a nine-country effort to identify cost-efficient pre- and post-harvest control options based on a multidisciplinary study of the farm-to-table pork production system. Data from the nine countries represent a range of production systems in the EU. Figure 4 shows the pig-meat production chain and the distinction between pre-harvest and post-harvest control. The epidemiological and diagnostic data were collected and evaluated from testing in the participating countries. Control options were identified, as shown in Table 2. Combined collection of data, information from previous studies and expert opinions were used to develop measures of effectiveness of the various control measures. In this assessment, irradiation was the only measure shown to bc 100 percent effective.

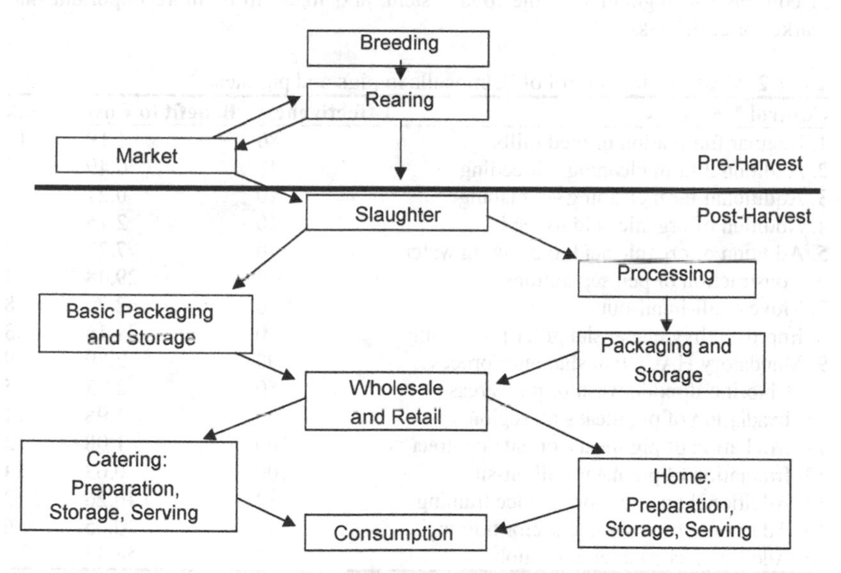

Figure 4. The pig-meat production chain. From Wong and Hald (2000), *Salmonella* in Pork

The results of the epidemiological and diagnostic data and risk assessment were used in the economic assessment of the alternative controls. Costs were developed based on costs assigned to the control measures; benefits were developed based on costs of illness averted, accounting for the control-measure effectiveness. The findings of the economic assessment show that consumer education, on-site food-service training, improved hygiene at slaughter/processing and a chlorine-dioxide wash of pig carcasses were among the controls with highest benefit–cost ratio. All of these measures occurred relatively close to the consumer and retail level. The HACCP controls did not rank among those with highest benefit–cost ratios. However, aspects of the plant-level production process that did rank highly (such as the sanitizing wash) may be related to tighter control of product and treatments at the slaughter/processing level.

Implications for identifying cost-efficient controls in farm-to-table assessments

There are four aspects of a farm-to-table risk, system-wide perspective that have implications for understanding how economics contributes to understanding likely costs of controls and interventions. First, the growing number of studies that take a system-wide, farm-to-table approach have provided good evidence on the weaknesses in data availability and knowledge needed to implement such studies. The significant demand for data and information at each stage is a challenge to each of the efforts and in some cases limits the overall ability to rank the full set of control options. Despite the limitations, the studies reviewed do highlight the contributions of economic information. The distribution of costs is likely to be an important aspect of the effect of controls and regulation in the food system, and likely to be more important than market-price effects.

Table 2. Measures for control of Salmonella in pigs and pig meat

Control Measures	Effectiveness	Benefit to Cost	Rank
1. Regular fumigation of feed mills	80	1.19	12
2. Additional farm cleaning – breeding	40	0.49	15
3. Additional farm cleaning – finishing	40	0.27	16
4. Addition of organic acid to feed	40	2.15	10
5. Addition of organic acid to drinking water	40	7.77	7
6. Construction of pen separations	20	29.48	4
7. Move to all-in/all-out	40	2.79	8
8. Improved hygiene at slaughter/processing	80	31.16	3
9. Mandatory HACCP at slaughter/process	90	2.79	9
10. Chlorine-dioxide wash of pig carcass	50	23.3	5
11. Irradiation of pig meat - all regional	100	1.98	11
12. Irradiation of pig meat - on-site/contract	100	1.08	13
13. Irradiation of pig meat - all on-site	100	0.63	14
14. Additional on-site food-service training	52	46.26	2
15. Additional food service sector training	52	10.65	6
16. Additional consumer education	17	54.44	1

From: Wong and Hald (2000), *Salmonella* in Pork

A second finding is that greater benefits are likely to be achieved and at lower cost to society, with incentive-based measures. Allowing market adjustments to mitigate costs, improving upon existing market incentives, and facilitating private market mechanisms that include contracting of product will be the most effective way to reduce the costs to society of food-borne diseases.

Third, regulation is likely to have an impact on long-run incentives to invest in new technologies or inputs. Not only do technologies related to control of hazards contribute to reducing hazard, but technologies designed to identify the cause of failure will limit the losses in the system associated with failures.

And finally, a risk-based systems approach can be the best way to understand the costs, incentives and risk outcomes resulting from alternative interventions. Because many food-safety problems stem from problems that are systemic in nature, analysis and policy prescriptions should also have a systemic orientation. Economists have addressed problems of systemic risks in other industries (such as banking and finance). However, they have much to learn from scientists and others who study the biological and physical nature of the system linkages. The combined tools will be needed to reduce the food-safety hazards in the food system with most efficient use of available resources.

References

Antle, J. M., 2000. No such thing as a free safe lunch: the cost of food safety regulation in the meat industry. *American Journal of Agricultural Economics,* 82 (2), 310-322.

Bunte, F. H. J., 1999. The vertical organization of food chains and health and safety efforts. *In:* Unnevehr, L. ed. *Economics of HACCP : new studies of costs and benefits. Proceedings of a NE-165 Conference.* Eagan Press, St Paul, MN, 285-300.

Cohen, J. T., Duggar, K., Gray, G. M., et al., 2001. *Evaluation of the potential for bovine spongiform encephhalopathy in the United States.* Available: [http://www.hcra.harvard.edu/pdf/madcow_report.pdf] (29 Jan 2002).

Crutchfield, S. R., Buzby, J. C., Roberts, T., et al., 1997. *An economic assessment of food safety regulations: the new approach to meat and poultry inspection.* Economic Research Service-USDA, Washington, DC. Agricultural Economic Report no. 755.

Egg safety from production to consumption: an action plan to reduce salmonella enteritidis illnesses due to eggs, 1999. Available: [http://www.foodsafety.gov/%7Efsg/ceggs.html] (29 Jan 2002).

Food labeling, safe handling statements, labeling of shell eggs; refrigeration of shell eggs held for retail distribution: final rule, 2000. *Federal Register, US Food and Drug Administration,* 65 (234), 76091-76114. [http://vm.cfsan.fda.gov/~lrd/fr001205.html]

Golan, E. H., Vogel, S. J., Frenzen, P. D., et al., 2000. *Tracing the costs and benefits of improvements in food safety: the case of hazard analysis and critical control point program for meat and poultry.* Economic Research Service-USDA, Washington, DC. Agricultural Economic Report no. 791.

Goodhue, R. E. and Rausser, G. C., 1999. Value differentiation in agriculture: driving forces and complementarities. *In:* Galizzi, G. and Venturi, L. eds. *Vertical relationships and coordination in the food system.* Physica-Verlag Publishers, Heidelberg, 93-112.

Grijspaardt-Vink, C., 1995. HACCP in the EU. *Food Technology,* 49 (3), 36.

Guidelines for preparing economic analysis, 2000. Available: [http://yosemite.epa.gov/ee/epa/eed.nsf/pages/guidelines] (6 Mar 2003).

Hathaway, S., 1995. Harmonization of international requirements under HACCP-based food control systems. *Food Control,* 6 (5), 267-276.

Hayes, D. J., Jensen, H. H., Backstrom, L., et al., 2001. Economic impact of a ban on the use of over the counter antibiotics in U.S. swine rations. *International Food and Agribusiness Management Review,* 4 (1), 81-97.

Hazard analysis and critical control point (HACCP): procedures for the safe and sanitary processing and improving of juice: final rule, 2001. *Federal Register, US Food and Drug Administration,* 66 (13), 6137-6202. [http://www.fda.gov/OHRMS/DOCKETS/98fr/011901d.pdf]

Hennessy, D., Roosen, J. and Jensen, H. H., 2002. *Systemic failure in the provision of safe food.* Center for Agricultural and Rural Development, Iowa State University, Ames, Iowa. CARD Working Paper no. 02-WP 299. [http://www.card.iastate.edu/publications/DBS/PDFFiles/02wp299.pdf]

Hennessy, D. A., Roosen, J. and Miranowski, J. A., 2001. Leadership and the provision of safe food. *American Journal of Agricultural Economics,* 83 (4), 862-874.

Henson, S. and Northen, J., 1998. Economic determinants of food safety controls in supply of retailer own-branded products in United Kingdom. *Agribusiness,* 14 (2), 113-126.

Jensen, H. H. and Unnevehr, L. J., 1999. HACCP in pork processing: costs and benefits. *In:* Unnevehr, L. J. ed. *Economics of HACCP : new studies of costs and benefits. Proceedings of a NE-165.* Eagan Press, St Paul, MN, 29-44.

Jensen, H. H., Unnevehr, L. J. and Gomez, M. I., 1998. Costs of improving food safety in the meat sector. *Journal of Agricultural and Applied Economics,* 30 (1), 83-94.

Just, R. E., Hueth, D. L. and Schmitz, A., 1982. *Applied welfare economics and public policy.* Prentice-Hall, Inc, Englewood Cliffs, NJ.

MacDonald, J. M. and Crutchfield, S., 1996. Modeling the costs of food safety regulation. *American Journal of Agricultural Economics,* 78 (5), 1285-1290.

Martin, S. A. and Anderson, D. W., 1999. HACCP adoption in the US food industry. *In:* Unnevehr, L. J. ed. *Economics of HACCP : new studies of costs and benefits. Proceedings of a NE-165 Conference in Washington, D.C., June 1516,1998.* Eagan Press, St Paul, MN, 15-28.

McDowell, R., Kaplan, S., Ahl, A., et al., 1995. Managing risks from foodborne microbial hazards. *In:* Roberts, T., Jensen, H. H. and Unnevehr, L. eds. *Tracking foodborne pathogens from farm to table: data needs to evaluate control options. Conference proceedings.* USDA, Economic Research Service, Washington, DC, 117-124. Miscellaneous Publication no. 1532.

Narrod, C. A., Malcolm, S. A., Ollinger, M., et al., 1999. *Pathogen reduction options in slaughterhouses and methods for evaluating their economic effectiveness.* Available: [http://agecon.lib.umn.edu/cgi-bin/pdf_view.pl?paperid=1786&ftype=.pdf] (29 Jan 2002).

Nauta, M. J., Evers, E. G., Takumi, K., et al., 2001. *Risk assessment of Shiga-toxin producing Escherichia coli O157 in steak tartare in the Netherlands.* Rijksinstituut voor Volksgezondheid en Milieu, Bilthoven. RIVM Report no. 257851003.

Pathogen reduction hazard analysis and critical control point (HACCP) systems: final rule, docket no. 93-016F, 1996. *Federal Register, USDA, Food Safety and Inspection Service,* 61 (144), 38805-38989.

Procedures for the safe and sanitary processing and importing of fish and fishery products: final rule, docket no. 93N-0195, 1995. *Federal Register, US Food and Drug Administration,* 60 (242), 65096-65202.

Roberts, T., Ahl, A. and McDowell, R., 1995. Risk assessment for foodborne microbial hazards. *In:* Roberts, T., Jensen, H. H. and Unnevehr, L. eds. *Tracking foodborne pathogens from farm to table: data needs to evaluate control options. Conference proceedings*. USDA, Economic Research Service, Washington, DC, 95-115. Miscellaneous Publication no. 1532.

Team, E. c. O. H. R. A., 2001. *Draft Risk Assessment of the Public Health Impact of Escherichia coli O157:H7 in Ground Beef.* Available: [http://www.fsis.usda.gov/OPPDE/rdad/FRPubs/00-023NReport.pdf].

Unnevehr, L. J. (ed.) 1999. *Economics of HACCP: new studies of costs and benefits. Proceedings of a NE-165 Conference*. Eagan Press, St Paul, MN.

Unnevehr, L. J., Gomez, M. I. and Garcia, P., 1998. The incidence of producer welfare losses from food safety regulation in the meat industry. *Review of Agricultural Economics,* 20 (1), 186-201.

Unnevehr, L. J. and Jensen, H. H., 1996. HACCP as a regulatory innovation to improve food safety in the meat industry. *American Journal of Agricultural Economics,* 78 (3), 764-769.

Unnevehr, L. J. and Jensen, H. H., 1999. The economic implications of using HACCP as a food safety regulatory standard. *Food Policy,* 24 (6), 625-635.

Unnevehr, L. J. and Jensen, H. H., 2001. Industry compliance costs: what would they look like in a risk-based integrated food system? *In:* Taylor, M. and Hoffman, S. eds. *Risk-based priority setting in an integrated food safety system. Resources for the Future conference on setting food safety priorities: toward a risk-based system, Washington, D.C., May 23-24, 2001*. Resources for the Future, Washington, DC.

White paper on food safety, 2000. Available: [http://europa.eu.int/comm/dgs/health_consumer/library/pub/pub06_en.pdf] (6 Mar 2003).

Whitehead, A. J. and Orriss, G., 1995. Food safety through HACCP. *Food, Nutrition and Agriculture,* 5, 25-28.

Wong, D. M. A. L. F. and Hald, T. (eds.), 2000. *Salmonella in Pork (SALINPORK): pre-harvest and harvest control options based on epidemiologic, diagnostic and economic research. Final report to the Commission of the European Communities, Agriculture and Fisheries (FAIR)*. The Royal Veterinary and Agricultural University, Department of Animal Health and Animal Science and the Danish Veterinary Laboratory, Danish Zoonosis Centre.

Ziggers, G. W., 1999. HACCP, vertical coordination and competitiveness in the food industry. *In:* Unnevehr, L. J. ed. *Economics of HACCP : new studies of costs and benefits. Proceedings of a NE-165 Conference*. Eagan Press, St Paul, MN.

---

[1] One of the difficulties with evaluating interventions to control pathogens is that their effectiveness is generally measured under laboratory conditions where samples are intentionally inoculated with high levels of pathogens. In meat processing plants, levels of contamination are low, and many more samples would be needed to assess the effectiveness of a technology.

8

The economics of HACCP: farm-to-table analysis

Mogens Lund[*]

Introduction

Risk analysis seems to be one of the basic principles for the food-safety policies in the EU. According to the EU White Paper the food policy should be based on the three elements in risk analysis, that is risk evaluations (scientific consultancy and data analysis), risk management (regulation and control) and communication of risk factors (*Food safety - a worldwide public health issue* 2000). However, it is remarkable that nothing explicitly is said about the role of economics in the EU White Paper on Food Safety. Does this reflect that the EU politicians and authorities reject to include economists in the risk-analysis process in considering the appropriate food policies to adopt and the policy regulations to deploy?

By arranging a workshop with the title "New Approaches to Food-Safety Economics" it is once again stressed that we need to readdress the role that economics might and can play in policies and strategies for food-safety improvement. One renewed approach might be to adopt a system methodology in order to ensure an enhanced integration of cost-benefit evaluations into risk analyses. In what follows, I will attempt to discuss the benefits and challenges of such a system methodology by using an ongoing Danish research project concerning the economics of food quality and safety as an example.

Background and aim of the Danish project

The project (Food quality and safety – Consumer behavior, food supply chains and economic consequences) was initiated in 2001 by the Danish Research Institute of Food Economics. The motivation for starting the project was a need to improve our general knowledge about quantitative relationships and behavioral parameters related to quality and safety attributes in Danish food products, not least seen under the changing policy regulations that Danish producers, manufacturers and consumers are facing.

The project has three major aims. Firstly, the purpose is to investigate consumers' attitudes towards food quality and safety, including the effects of labeling, marketing and product identity for consumer behavior. Secondly, the aim is to quantify and evaluate the consequences of changing consumer requirements for the downstream food chain. This will be accomplished by case studies where selected food products are analysed and central issues are identified through the different stages in the chain, i.e. from the consumer via retailers and distribution to processing and primary production. Thirdly, by using generated price and cost information holistic scenario

[*] *Farm Management and Production Systems Division, Danish Research Institute of Food Economics, Rolighedsevej 25, 1958 Frederiksberg, Denmark*

A.G. J. Velthuis et al. (eds), New Approaches to Food-Safety Economics, 79-87.

analyses are carried out in order to evaluate the economic perspectives for production and marketing of food products with specific safety and quality attributes.

The research efforts in the project are mainly directed towards executive managers in the food industry and policy regulators, who are all involved in strategic decision-making concerning the formulation of future quality and safety strategies related to the production and distribution of food products to Danish and foreign consumers.

The contents of the project

The project will be carried out in five subprojects as shown in figure 1:

I. Establishment of the project
II. Consumer behavior and marketing
III. Quality and safety in the supply chain
IV. Holistic scenario analyses
V. Recommendations and perspectives.

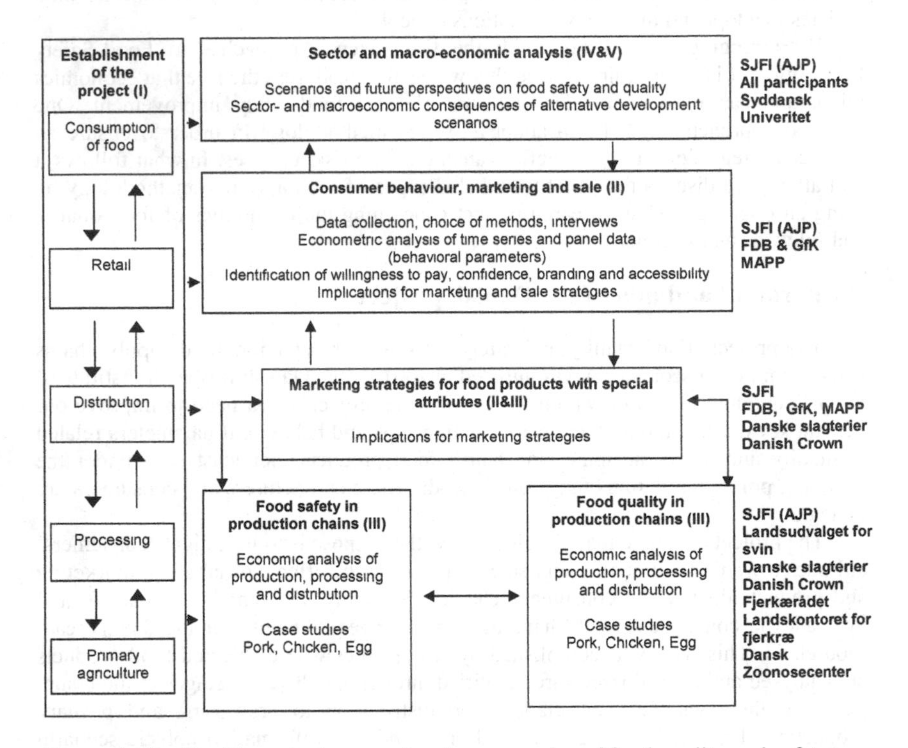

Figure 1. The Danish project concerning the economics of food quality and safety

I. Establishment of the project

The project is designed as a cooperative project between the Danish Research Institute of Food Economics (SJFI) and a range of other institutions, including a data-survey institute (GFK), a center for research on customer relations in the food sector

(MAPP), a surveillance center for zoonosis (Dansk Zoonosecenter), an institute for human-health economics (Syddansk Universitet), a major food retailer (FDB), the biggest Danish slaughterhouse (Danish Crown) and several development and extension associations (Danske Slagterier, Landsudvalget for Svin, Fjerkrærådet and Landskontoret for fjerkræ).

The aim of subproject I is to establish a general conceptual framework for the total project and to identify the working tasks of the individual participants and their collaborative relationships.

II. Consumer behavior and marketing

The aim of the subproject is to perform quantitative analyses on the behavior of Danish consumers with respect to the demand of food products. The analyses include the division of the food consumption into different product groups and different categories of quality and safety as well as estimates of the sensitivity of the consumption patterns with respect to price and income changes for different product groups and levels of quality. The subproject includes four parts:

- Establishment of a database for consumer data and characteristics, food prices, etc.
- Estimation of parameters for consumer behavior
- Analyses of the impact of information asymmetries, credence, accessibility, etc.
- Evaluation of alternative marketing strategies.
- Econometric methods are employed to estimate consumer behavior and willingness to pay for high-quality foods by formulating alternative demand systems that will include variables such as income, prices and safety and/or quality indicators.

III. Quality and safety in the supply chain

The purpose of the subproject is to identify and evaluate strong and weak points in existing supply chains and on this basis specify minimum standards for safe food products. A second objective is to estimate the costs associated with the production, distribution and marketing of specific food products characterized by high levels of safety and quality. As indicated in the figure, analyses will be performed for three case studies, including pork, eggs and chickens. These three food products should represent different degrees of integration in the food supply chains and different strategies to comply with food-safety regulations.

Industrial organization theory and new institutional economics will be applied to evaluate the performance of the existing supply chains, while the best methods to estimate the costs in the different stages of the chains yet have to be decided.

II & III Evaluation of marketing strategies

Based on the knowledge generated in subprojects II and III, a number of different marketing strategies will be evaluated under due consideration to, for instance, the design of relevant labeling schemes. Thus, the aim of the analysis is to identify the opportunities and limitations of alternative marketing strategies for food products with specific quality and safety characteristics such as *Salmonella*-free products. Organizationally, the analysis of the marketing strategies is considered as a part of subproject II.

IV. Holistic scenario analyses

The aim of this subproject is to integrate the results from the other subprojects into holistic economic analyses. Thus, to estimate the total benefits and costs related to an improved level of food quality and safety and to evaluate the future economic perspectives for the production and marketing of Danish food products with enhanced quality and safety. The subproject contains three parts:

– Overall social evaluations related to food-borne diseases
– Construction and evaluation of a baseline scenario for the future development of the Danish food industry
– Construction and evaluation of alternative scenarios for the future development of regulatory polices, changing consumer preferences etc.

The scenario evaluations will be performed by use of a Danish general equilibrium (CGE) model with the nickname "Aage", which by construction can simulate the interactions between different economic sectors and markets in the Danish economy. However, in the project no considerations are explicitly taken into account with regard to the international trade effects of potential changes of the Danish and/or the EU food policies.

V. Recommendations and perspectives

The objective of this latter subproject is to summarize the results and experiences from the four subprojects and to formulate policy recommendations in relation to the:

– marketing strategies of the food industry
– formulation of future food policies
– new research and development activities related to food quality and safety.

By using this project as an example, I will attempt to address the key questions as formulated in the workshop program under the headline "Farm-to-table Risk Analysis and HACCP".

How can cost-benefit analysis be integrated into farm-to-table risk analysis?

In a framework as shown in Figure 1 it is in principle possible to integrate cost–benefit analyses into farm-to-table risk analysis. Since, food economics is all about determining the point where the marginal costs spent on e.g. food safety just balance the marginal benefits gained from the resultant increase in food safety. Such economically optimal solutions also imply that no change in reallocation of resources from e.g. environmental protection to food safety could increase overall human welfare.

In practice, the integration is of course much more difficult. One major problem we are facing is the problem of valuing the benefits of an increase in food safety. These benefits are normally assumed to be the reduction in suffering and an improvement in life expectancy, which are intangible and uncertain by nature and therefore difficult to measure. However, when governments for example decide to construct new roads rather than building a new hospital, they are also implicitly valuing human life.

In my opinion one major challenge in order to integrate cost–benefit analysis into farm-to-table risk analysis is to establish a much better cooperation between researchers from the natural and social sciences and also between research communities, private companies and the public authorities. For example, in recent years animal diseases and production losses due to these diseases have been investigated to a greater extent. However, most of this information has not been utilized in a holistic manner to evaluate the total impact on animal production and on the related industries. The experiences from our Danish project are that it requires a lot of time and effort to establish and maintain such types of networks.

What lessons can we learn from modeling animal-disease risks?

Obtained experiences indicate that a number of important lessons can be learned from modeling animal-disease risks. First, we know that traditionally decisions on control strategies at the herd level have been based on the farmer's or herd advisor's subjective judgement, intuition, experience, attitudes to risk and presumed economic costs and benefits of control strategies. Therefore, we should not expect practical decision-makers to make important decisions on the basis of cost–benefit studies alone. The challenge is of course to prove that the economist's framework to think and make analyses can and should be an important contribution to the decision-making process (Houe et al. 2001).

Secondly, experiences with decision-support systems in animal-health management show how difficult it is to construct real integrated models for practical decision-making. We normally agree that biological and economic modeling should be integrated in order to provide improved decision support to farmers, but often this is not the case. Frequently, the economic model is just an appendix to the technical/biological model; the reason may be that the model builders from natural sciences and social sciences have no common training, experiences or professional working fields (McInerney 2001).

Thirdly, we have learned how difficult it is to obtain knowledge about all risk factors and their causal relationships, and how difficult it is to gather sufficient and reliable data on the effects of these risk factors.

Fourthly, it is increasingly realized that modeling animal diseases should be done in a wider context. As noted by Dijkhuisen (Dijkhuisen 1998), a more integrated chain approach is necessary in order to provide fast and reliable tracing of animal diseases to meet safety requirements. A system perspective seems to be appropriate as farms are consolidated into fewer, but larger production units that are operating with an industrial mentality, which creates the opportunity for supplying differentiated agricultural products into the food chain (Downey 1996).

How does a system approach differ from examination of stages of the food chain separately?

Food products are produced and consumed through a series of stages from farm production to the end consumer as illustrated in Figure 1. Understanding the competitiveness of all the stages is important, not only in terms of determining how the total chain performs, but also in understanding the appropriate role of e.g. public regulation. Most economic studies, however, have only considered selected parts of the food supply chain.

The term "system" might better describe the business challenges faced by food and agribusiness firms than the term "chain". The term "chain" is associated with a linear understanding that is insufficient to understand really all the interactive and dynamic business games that are increasingly taking place among firms in the battle for gaining competitive advantages and control in the food markets. Furthermore, food supply chains often involve many interconnected stages and inputs, and failure at any point in the chain may require the product to be condemned. Several examples of serious food contamination by pathogens in different countries have illustrated these interconnections, and high costs associated with such food scares have been reported in the literature. Thus, it seems that the most serious food-safety problems can be characterized as systemic by nature. According to Hennessy, Roosen and Miranowski (Hennessy, Roosen and Miranowski 2001), two of the most important aspects of systemic failure are interconnectivity, which provides the technological potential for systemic failure, and the incentive problems, which provide the economic potential for systemic failure.

Measures to prevent food-safety problems from recurring are more likely to be successful if these interconnectedness and incentive aspects are understood from a systemic perspective. One example might be the understanding the total effect of supply management (Downey 1996). Another highly relevant example might be an improved understanding of the investment incentives in the chain and the problems concerning the division of the economic surplus among the actors in the supply chain. Thus, a systemic approach may have the potential to avoid sub-optimal solutions.

Does a system approach provide lower cost solutions?

At least, by adopting a system approach it is easily recognized that different measures can be taken to improve the safety of the food supply. Many regulatory alternatives are available along the chain from the farm to the table that might implement changes in production, processing, distribution and consumption in order to reduce the health risks associated with food-borne diseases. Among these are (Crutchfield et al. 1997):

- Improving the meat and poultry inspection system;
- Educating consumers, retailers, and food-service workers, and promoting safe food handling;
- Irradiating meat and poultry products; and
- Using market-oriented approaches to food safety: labeling, branding, legal incentives, and providing food-safety information about products or production methods.

All such options could help improve the safety of meat, poultry and other food products, but to make optimal choices among these alternatives requires estimation of the costs and benefits of each policy followed by a ranking of the available alternatives. Some sort of a system approach will also be necessary in order to evaluate all the distributional consequences of implementing new regulatory policies for farmers, food processors, retailers and consumers.

What will be the implication of the proposed HACCP directive from the EU that will apply to most of the food chain?

Governments, public agencies and international organizations (WHO, FAO) are increasingly recognizing HACCP as one of the most effective means to provide safe food. The basic idea of HACCP systems is to keep control by identifying the critical control points (CCPs) in the food production and distribution processes that are most important to monitor (Mortimore and Wallace 2001). As noted by Unnevehr and Jensen (Unnevehr and Jensen 1999) there is, however, some disagreement concerning the consequences of mandatory imposition of HACCP. Among other issues they mention disagreement regarding how effectively HACCP will be to control or eliminate some food-safety hazards, the controversy regarding whether it improves or reduces regulatory oversight and whether it allows firms to meet food-safety objectives in the most efficient manner or it is overly prescriptive.

Of course, it is possible to implement HACCP systems in the whole food chain, although HACCP is originally developed as a food-safety management tool to be used in individual firms. Currently it is probably most widespread in the food-processing industry. If the HACCP principles have to be implemented throughout the whole food chain, the requirements will be different and all the direct and indirect consequences will be difficult to predict. All the potential hazards in the chain should be identified and all scientific information needed for systemic risk assessments should be provided. It also raises the question about who should set the food-safety standards and critical limits of the CCPs and who is assumed to carry out all audit and verification activities.

If HACCP is applied to manage food safety in every stage of the food system from farm to table, we furthermore have to investigate if the total system is designed to provide enough feedback to make appropriate corrective actions. One of the characteristics of the response to food-safety issues is that it should be rapid because food safety may be an emotive issue among consumers, and when it comes to real food scares, HACCP systems seem to be of limited value, because the system works by negative-feedback mechanisms. In dealing with food scares there may instead be a need for crisis management, which requires use of other mechanisms such as "crisis teams".

There may also be some concern that mandatory application of HACCP in the whole food chain will make it more difficult to coordinate public regulation with private incentives for using quality-control systems. Private systems include self-regulation and various forms of certification by other parties (Henson and Caswell 1999).

How can we measure the costs and benefits of alternative interventions or systems of prevention?

The main problem with measuring the costs and benefits of such regulations is that food safety itself is very difficult to measure. This is also the case in our Danish project, where we have to estimate both the benefits and costs of alternative safety regulations in the food chains. As noted by Antle (Antle 1999), information about the various quality and safety attributes of food products is not perfectly known by consumers, producers, government regulators and researchers. Microbial pathogens cannot easily be identified in the production process and their health effects are often difficult to recognize. This is reflected in the reported empirical cost–benefit analyses, which indicate that the existing data can only provide us with highly uncertain estimates of the benefits and costs associated with new food-safety regulations (Crutchfield et al. 1997).

As already noted, the benefits of food-safety regulation are reductions in risks of suffering and mortality associated with contaminated foods. A number of methods have been developed to measure health risks and risks of dying. The most simple method to value health risks is the cost-of-illness method (Kenkel 1994), but although it is simple to use, it lacks a theoretical foundation. A more proper theoretical approach is adaptation of the willingness-to-pay methods, which will be used in the

Danish project. There seems, however, to be some controversy about their validity and their ability to produce sufficiently general results (Antle 1999).

There are three well-known approaches available to estimate regulatory costs. The first one is the accounting approach, which is rather straightforward to use, but also has a number of limitations. The most important is that it is not possible to measure the overall efficiency of any changes in input variables such as quality control. The second is the economic-engineering approach by which it is possible to get a rather detailed description of the cost structure and whereby it is also in principle possible to derive the cost function. The economic-engineering approach is, however, rather time-consuming to apply. The last approach is to estimate cost functions by use of econometric methods. Traditional econometric estimation methods generally require greater data sets compared to the two other methods, but the advantages of the econometric methods are that it is possible to study actual cost behavior and that statistical tests concerning the underlying production structure and behavior can be carried out.

Final remarks

I have no doubt that much can be gained from adopting a systemic perspective in cost-benefit assessments of food-safety regulations, but I do not think that it is possible to point out the best economic tools. The choice of methods will always be dependent on the specific circumstances. But a general future research challenge in measuring the benefits and costs of food-safety regulations is – in my opinion – to develop methods that can make the best use of the available data, which are limited and imperfect (Antle 1996; Antle 1999). Another important potential research area is to study the implementation costs associated with new forms of regulations. Depending on the time frame given to implement new regulations, firms will have more or fewer opportunities to learn how to handle new routines, and these may have important impact on their adjustment costs.

References

Antle, J. M., 1996. Efficient food safety regulations in the food manufacturing sector. *American Journal of Agricultural Economics,* 78, 1242-1247.

Antle, J. M., 1999. Benefits and costs of food safety regulation. *Food Policy,* 24 (6), 605-623.

Crutchfield, S. R., Buzby, J. C., Roberts, T., et al., 1997. *An economic assessment of food safety regulations: the new approach to meat and poultry inspection.* Economic Research Service-USDA, Washington, DC. Agricultural Economic Report no. 755.

Dijkhuisen, A. A., 1998. Animal health and quality management in the livestock production chain: implications for international competitiveness. *In:* Zuurbier, P. J. P. ed. *Proceedings of the third international conference on chain management in agribusiness and the food industry.* Wageningen Agricultural University, Wageningen, 31-43.

Downey, W. D., 1996. The challenge of food and agri products supply chains. *In:* Trienekens, J. H. and Zuurbier, P. J. P. eds. *Proceedings of the 2nd international conference on chain management in agri- and food business.* Wageningen Agricultural University, Wageningen, 3-13.

Food safety - a worldwide public health issue, 2000. Available: [http://www.who.int/fsf/fctshtfs.htm] (6 Mar 2003).
Hennessy, D. A., Roosen, J. and Miranowski, J. A., 2001. Leadership and the provision of safe food. *American Journal of Agricultural Economics,* 83 (4), 862-874.
Henson, S. and Caswell, J., 1999. Food safety regulation: an overview of contemporary issues. *Food Policy,* 24 (6), 589-603.
Houe, H., Soerensen, J. T., Otto, L., et al. (eds.), 2001. *Animal health economics. Proceedings of workshop held by Research Centre for the Management of Animal production and Health, Danish Institute of Agricultural Science, Foulum.* Danish Institute of Agricultural Science, Foulum. CEPROS report no. 7.
Kenkel, D., 1994. Cost of illness approach. *In:* Tolley, G., Kenkel, D. and Fabian, R. eds. *Health values for policy: an economic approach.* University of Chicago Press, Chicago.
McInerney, J., 2001. Conceptual consideration when developing decision support tools for herd health management. *In:* Houe, H., Soerensen, J. T., Otto, L., et al. eds. *Animal health economics. Proceedings of workshop held by Research Centre for the Management of Animal production and Health, Danish Institute of Agricultural Science, Foulum.* Danish Institute of Agricultural Science, Foulum. CEPROS report no. 7.
Mortimore, S. and Wallace, C., 2001. *HACCP.* Blackwell Science, Malden, MA.
Unnevehr, L. J. and Jensen, H. H., 1999. The economic implications of using HACCP as a food safety regulatory standard. *Food Policy,* 24 (6), 625-635.

TRANSPARENCY IN INTRA-EU AND INTERNATIONAL TRADE

9

Regulating food safety in the European Union

Stephan Marette[*]*, Jean-Christophe Bureau, Bénédicte Coestier and Estelle Gozlan*

Introduction

Recent and highly publicized incidents in the European Union have urged policy makers to consider changes in the food-safety regulations affecting domestic and imported food products. The Bovine Spongiform Encephalopathy (BSE) crisis, followed by a series of more anecdotal food-contamination cases, has triggered a major change by placing food safety on the top of the agenda of most EU governments.

The BSE crisis has had dramatic economic consequences, in addition to a significant number of deaths (most of them in the United Kingdom). After the finding of a possible link between BSE and a new strain of human disease, demand for beef fell and export bans hurt the entire sector throughout the European Union, costing billions of euros. The poor management of this crisis by British and other national and European authorities has also led to less visible, but very large impacts in terms of citizens' emotion. At a smaller scale, the 1999 contamination of feed in Belgium also had serious trade impacts. When it became public that some fat used in animal feed was inadvertently contaminated with cancer-inducing dioxin, some animal imports from several European countries were banned in a number of regions, including the United States. This resulted in a decline in meat production in Belgium, hitting particularly the swine and poultry sectors. Again, a major effect was the loss of consumer confidence, which at some point led to imaginary risks. The withdrawal of Belgian soda (Coca Cola) from the market in 1999, in spite of any scientific evidence, can be seen as an indirect consequence of the dioxin crisis. The frequent food scares that have followed the release of information about pathogens such as *Listeria* found in some prepared meat and some soft cheese during the years 2000 in France, can also be attributed to a general mistrust of consumers (statistics show that the number of food-borne diseases due to *Listeria* has actually been falling and that fatal casualties are very exceptional). Such food scares seem to be mainly due to the release of information, which did not occur in the past.

A particular situation in the EU is also the growing mistrust in science over the last decade. France is a typical example where the government has minimized the effect of major accidents, which have fueled suspicion and eroded public confidence. (The importance of asbestos-related cancers has been largely hidden under the pressure of the industry, and, when disclosed, past responsibility of mandated doctors in spreading wrong information has had a very negative effect on public opinion.) Involvement of scientists in hiding information from the public in the nuclear sector

[*] *UMR Economie Publique INRA, 16 rue Claude Bernard, 75005 Paris, France*

A.G. J. Velthuis et al. (eds), New Approaches to Food-Safety Economics, 91-106.

has had a similar effect. (Scientists from government agencies claimed that the Chernobyl radioactive cloud has stopped exactly at the French border, something that nobody has actually believed.) So has the continuous denying of the involvement of public agencies in spreading HIV-contaminated blood, until journalists disclosed evidence. Poor management of information and assurances from government-appointed scientists made the mistrust of science a very sensitive issue in the food sector. New production methods driven by technology have added to consumer unease. Consumer concern about genetically modified organisms and growth activators cannot be understood without taking this background into account. Indeed, European consumers have a peculiar position regarding biotech food. While GMOs seem to have been tacitly accepted in the US, they have caused large protests in Europe. However, environmental protection organizations and US consumers play a dynamic role in the Starlink case (see Taylor and Tick (Taylor and Tick 2001)). In France, GMO experiments have led to numerous acts of destruction (and to penalties for the activists), up to a point where experimental research is practically impossible on the territory. While the reluctance of consumers to purchase processed food containing GM ingredients is perhaps overestimated, the obligation of labeling has led processors to avoid using GM material whenever they could.

The BSE, dioxin and other food crises have increased the demand for more regulation and stricter enforcement of existing rules across EU countries. While these incidents acted as catalysts, the background was already a demand for more priority given to food safety on the policy agenda of the EU. As incomes rise, consumers become more demanding and are more prepared to pay for a regulatory regime that provides higher standards and minimizes risks.

Regulations are also evolving because of the international environment. International agreements have urged the EU to implement new instruments and new regulatory procedures. The Uruguay Round, which resulted in a constraining legal framework, with the creation of the Dispute-settlement body of the World Trade Organization (WTO), shapes the regulatory framework in the food-safety area. Disputes falling under the Sanitary and Phytosanitary (SPS) agreement have led the EU to implement measures that were previously unfamiliar, including systematic risk analysis, in the sense defined under the *Codex Alimentarius* measures (see Box 1). It is noteworthy, for example, that the 1997 arbitration of the WTO dispute on hormone-treated beef had taken place while EU governments were relatively unprepared to justify their measures by such analysis, which subsequently became routine procedure.

More generally, the Uruguay Round agreements have changed dramatically the way countries regulate food safety. Since 1995 (enforcement of the Marrakech Agreement), all WTO members have had to comply with the TBT Agreement, and a country may not reject panel conclusions simply because they are unfavorable. Many examples may be given of countries that have changed their restrictive practices and the very large number of notifications and discussions on potential flashpoints have made it possible to solve problems without embarking on the dispute-settlement procedure under the WTO.

The economic consequences of adopting certain standards, in particular in the *Codex Alimentarius*, have become very important, since these standards can now provide a basis for justifying or challenging an import ban. A 1997 WTO panel concluded to the non-conformity of the EU ban on hormone-treated beef, because most of the hormones incriminated had been tested within the *Codex*, which had set maximum residue limits.

Box 1: The international framework for food-safety regulations

Various publications from US and EU government agencies show that domestic regulations impede imports in almost all countries. Regulatory barriers in the European Union are often pointed out by US agencies (Bureau, Gozlan and Marette 1999). The EU ban on hormone-treated meat is one of the most quoted examples. In the EU Commission's market-access database, the pages relative to Japan are particularly impressive. Even Australia, a country known for low tariffs, has technical standards, which often preclude imports. The US conditions of sanitary inspection, with long and unpredictable delays, open lists for insects, which make import authorizations unpredictable, and complex quarantine rules, are also accused of making it unnecessarily difficult to export food products to the United States.

As a result of multilateral negotiations, rules have been introduced which aimed at minimizing the negative effects of sanitary, phytosanitary and technical regulations on international trade. Before 1995, GATT panel decisions have established the general principle that international rules do not permit WTO members to restrict the imports of products on the basis of how they are produced. The Uruguay Round provides a framework for solving disputes, through the WTO's Dispute Settlement Body. It tackles the problem of non-tariff trade barriers through the Sanitary and Phytosanitary (SPS) agreement and a strengthened Technical Barriers to Trade (TBT) agreement. And it gives greater importance to international bodies, especially Codex Alimentarius, an international code of standards for human-health protection under the auspice of the Food and Agriculture Organization (FAO) and the World Health Organization (WHO). The SPS Agreement covers health risks (food safety) arising from additives, contaminants, toxins and pathogens contained in food products. Members' measures that are based on international standards are deemed to be in accordance with the SPS Agreement. (When a regulation complies with an international standard, there is no need to notify the WTO or to justify it against a challenge from another State.) Members may introduce or maintain SPS measures that result in a higher level of protection than that achieved by the relevant international standards, if there is scientific justification or if it is a consequence of a level of SPS protection deemed appropriate by the Member based on an appropriate assessment of risks. The agreement asserts the right of signatory countries to take the measures they deem necessary to protect human, animal or plant life or health, provided that such measures are based on scientific principles, are not maintained without sufficient scientific evidence, and are not applied in an arbitrary or unjustifiable way. The agreement states that sanitary measures may not be used for protectionist purposes. The SPS agreement encourages the harmonization of SPS measures based on internationally accepted standards, guidelines or recommendations. It also encourages the signatories to conclude bilateral and multilateral agreements on the recognition of equivalent sanitary and phytosanitary measures of other WTO members.

The scope of the 1979 Technical Barriers to Trade (TBT) agreement was also extended during the Uruguay Round. Compliance with relevant international standards is encouraged. The TBT agreement is wide-ranging and covers all technical regulations and standards except those falling under the SPS agreement, including those relating to packaging and labeling.

The Uruguay Round also resulted in the introduction of a notification procedure, which acts as an early-warning system when national TBT or SPS regulations would be liable to restrict trade. If the disagreement persists, the WTO may, at the request of the parties involved, set up a panel to examine the problem. Final arbitration can lead to compensation measures authorized by the WTO, as the ones taken by the United States against the European Union, which refused to amend its legislation after being recognized as violating the SPS agreement with its import ban on hormone-treated beef.

The outcome of the panel, even though it was partially changed by the Appellate body in 1998, has changed completely the status of the international standards in the EU regulatory process. Because these standards are now at the core of economic disputes, "scientific" discussions have been difficult within the *Codex*. Up to now, the adoption of controversial standards with large economic consequences has been delayed (case of maximum limits on residues of rBGH or somatotropin, case of GMOs), or has been adopted with an ambiguous wording (standards on pasteurization of dairy products). However, in very complex and controversial cases such as GMOs, for example, either the adoption of precise standards, or the clarification of the status of GMOs by a WTO panel or appellate body in the international framework might be necessary in the future. They are likely to act as major constraints for the EU domestic regulation.

Changing the EU regulation on food safety

Among the many emerging public concerns, agro-food is one of the most visible sectors, and pressures for government intervention are high. The combination of public concern, the media coverage of the various crises, and the international framework, has led to considerable institutional changes in the EU. In Germany and the United Kingdom, the ministry of agriculture has now become a more general ministry, whose name indicates that farmers come second to consumer protection. France and the UK have established a new food agency with a broad mandate for health, safety and inspection responsibilities. The European Commission issued a White Paper after a wide debate on food policy and food law in January 2000 in order to restore consumer confidence (*White paper on food safety* 2000). It underlines the need for the establishment of the European Food Authority, whose creation was subsequently decided but whose localization still needs to be defined. These EU changes have also put food safety on the top of the agenda for accessing countries. Candidate countries have drawn "food-safety strategies" that outline the transposition and implementation of existing EU legislation. They are also supposed to draw on the White Paper in order to prepare some initiatives in this area (the implementation of the EU legislation in some candidate countries has already received approval from the Commission).

The White Paper proposes a wide range of measures to improve the corpus of legislation covering all aspects of food products as well as a new legal framework that covers the whole of the food chain. The objective is to establish a high level of consumer health protection and to attribute primary responsibility for safe food to the industry, producers and suppliers.

In particular, the White Paper stresses two areas of public regulation. The first one is the food-safety control, with control procedures at both national and European level. The paper acknowledges the need to create a coherent and transparent set of food-safety rules, and to recast the different control requirements, making all parts of the food production chain subject to official controls. The proposal is to improve, and to make more coherent, the system of national inspection services and controls. Controls at the borders of the EU would also be extended. The ability to trace products through the whole food chain is a key issue.

The second area of intervention is consumer information. The incentive is set on risk communication, and on providing consumers with essential and accurate information so that they can make informed choices. The European Food Agency will play a major role in the operations of risk assessment and management, but also in communication. The White Paper mentions binding labeling rules, through the codification of a labeling directive, including on novel food and on supplements, and involving more general rules on advertising messages. For instance, a Labeling Directive is proposed that would remove the current possibility not to indicate the components of compound ingredients that form less than 25% of the final product. The labeling provisions of the Directives on GMOs also show that a large role is left for "informational" instruments (see Box 2). One of the objectives is to ensure proper and consistent compliance and enforcement and to "avoid unnecessary administrative procedures".

Box 2: Regulation of GMOs in Europe

The body of regulation relative to GMOs relies on EU directives that have been adopted over the last ten years. The EU directive 90/220/CEE relative to the dissemination of GMOs is oriented towards the regulation of the protection of the environment. The main purpose of this directive is to regulate and harmonize the administrative procedures and the evaluations of GMOs for dissemination. This regulation has recently been amended by Directive 2001/18/CE in the spirit of the precautionary principle. The revised directive specifies the conditions that have to be fulfilled – assessment of the risks to the environment, a plan for monitoring in order to identify effects of GMOs on human health or the environment, among others – for the release of GMOs to proceed. The same conditions apply for GMOs or products containing GMOs that are placed on the market with additional proposals for labeling and packaging.

With respect to the issue of liability, the revised directive does not include any liability regime that would oblige marketers of GM food to give financial compensation for sanitary or environmental damage. However, article 32 states that the Commission should make a legislative proposal for implementing the Carthagena Protocol on biosafety; and article 27 of the Protocol states that the parties must adopt rules and procedures appropriate to responsibility and compensation for possible damage resulting from cross-border transport of live modified organisms.

The food-safety aspects of GMOs are regulated by the Novel Foods regulation CE 258/97. This regulation states that the safety of the products must be assessed when the new product contains a GMO or, resulting from genetic engineering, is substantially different from a traditional ingredient. In such cases labeling is required. The procedures to evaluate innocuousness of novel foods are described. Notice that the most worldwide used GMOs, Monsanto Roundup Ready soybean and Novartis Bt corn, were marketed before the regulation. A regulation that states that food containing these ingredients must be labeled was adopted in 1998 (1139/98). The regulation 49/2000 sets the threshold for mandatory labeling at 1% GMO, in order to account for accidental contamination. And the regulation 50/2000 provides for specific additional labeling requirements for food and food ingredients containing additives and/or flavorings that have been genetically modified or have been produced from genetically modified organisms.

The EU regulation raises concern regarding compatibility with international rules. The US-Canadian approach for evaluating GMOs (based on risk assessment) differs from the EU conception, which is more based on a precautionary approach to risk assessment and management. Up to now, the diverging conceptions have not led to an official dispute brought to the WTO. (Note that there has been a formal WTO notification of a dispute of Egypt challenging imports of tuna in GM soybean oil from Thailand (WT/DS205/1, WTO).) The odds that such a dispute may occur in the future are high. (Sheldon (Sheldon 2002) provides an analysis of the reasons for such a dispute along with a discussion of whether such a dispute can be resolved through existing WTO procedures.) The IATRC (*Agriculture in the WTO: the role of product attributes in the agricultural negotiations* 2001) has examined the issue; the outcome is uncertain. Given the various GATT agreements, the EU mandatory labeling could be challenged under either the SPS or the TBT agreement. The Article III of the GATT states that countries cannot discriminate between like goods on the basis of origin, raising the issue of the equivalence of GM and non-GM goods. The SPS-agreement approach of precaution (Article 5.7) is more restricted than the one underlying the EU regulation. The GATT provisions for ethical concerns (article XX) are unlikely to legitimate a ban. The EU could argue that there is a difference between the concept of equivalence and the concept of substantial equivalence mentioned in the agreements, and that GM crops are not equivalent to their GM counterparts and should therefore be labeled.

Liability or, more generally, self-enforcing procedures, have received less attention from the EU authorities than mandatory regulations and information-based procedures for ensuring food safety. The clarification of responsibilities of producers in the animal-feed sector can perhaps fall in the "liability" category, if we consider the various economic instruments. However, it also involves many mandatory regulation aspects (e.g. the revision of the legislation on animal-feed processing). More recently, in the discussions in the EU parliament on the regulation of GMOs, the proposals of the Commission for more liability of producers have been largely turned down by members of the European Parliament, a decision that environmentalists and consumer organizations have found difficult to accept.

The food crises that are mentioned in the White Paper in order to justify the need for a new approach of food safety, have not only generated a demand from citizens for more precaution, but also for more responsibility. In particular, the BSE crisis, as well as the foot-and-mouth disease (FMD) outbreak in 2000 has raised many questions about who should bear the costs of past mistakes. Up to now, it is mainly governments that have covered the costs, at least those costs that have been covered, given that many stakeholders have borne uninsured costs (including offal retailers, for example in the case of BSE, or the British rural tourism industry in the case of FMD. According to PricewaterhouseCoopers, the losses expected in the tourism industry from FMD crisis range from £1 billion (twice the loss of the agricultural sector) to £3.4 billion (The Economist, March 31st 2001)). Because of these food crises, beyond the claims of consumers for more safety, taxpayers are also demanding that producers, and possibly regulators, are more liable for possible contamination or mismanagement. This adds to the importance of more investigation on liability rules as a food-safety instrument.

Transatlantic difference about the role of liability

The role given to liability in the White Paper raises the issue of one of the most striking differences between the EU and US approaches. Punitive damages in product-liability action are very different in the United States and in European countries. In the United States, *ex post* liability clearly plays an important role in deterring firms from marketing unsafe products. Because of the potential outcome of tort law, firms often set up standards that exceed those required for passing the government approval process. Antle (Antle 1995) shows that this reduces the need for a "command and control" type of government intervention. In some EU countries, economic sanctions are very limited in the case of food-safety problems: when an unsafe product is marketed, resulting in the death of consumers, this rather results in penal sanctions for the manager than in large economic sanctions for the firm. Fundamental differences in the legal systems for protecting consumers from health hazards provide some justification for diverging conceptions on the role of government in setting standards. More generally, differences in the legal environment, such as *ex ante* regulation versus *ex post* litigation as a basis for legislation, contribute to explaining differences in governmental standards between countries.

If liability procedures are not seen as a central instrument in the White Paper, this does not mean that they are absent from the EU legislation on food safety. The Directive concerning liability for defective products (85/374/EEC, amended by Directive 99/34/EC) has required all member states to issue conforming "strict liability" laws on producers of defective products that cause personal injury or property damage (see Box 3).

Box 3: Liability for food products in practice : the case of France

The EU directive on the responsibility for defective products (85-374), has been translated into French law (98-389). This law introduces a large responsibility for producers, importers and sellers.

Traditionally, the French law, based on the Napoleonic code, emphasizes penal law. The French Penal Code specifies penal sanctions for endangering deliberately other people's life and health. In the Civil Code, though, the concept of civil responsibility applies for products one would legitimately expect to be safe. Absolute safety is not required but is an objective. The plaintiff must prove the damage, default and the causality between the default and damage. The producer can be responsible for the defect even though the product was made in respect of legal standards and good practices, unless the producer shows that he had not put the product in circulation, or that the defect did not exist when the product was put on the market.

The producer is not responsible if the state of knowledge or techniques do not make it possible to detect a default. However, there is an obligation for the producer to anticipate damages if there is suspicion, and to gather scientific information. That is, the "development risk" is not a complete exoneration of responsibility. However, lawyers acknowledge that the practical consequences of this 1998 law create many problems, since it opens the door for endless responsibility. The fact that appeal courts sometimes go back to the EU Directive in the interpretation of the law has also increased the scope for potential responsibility of producers even in the presence of development risk. The industry sees these changes in the legal system as a drift towards "compassionate law" that aims mainly at compensating a victim, even though the responsibility of the producer is questionable, and an obstacle to innovation.

A recent application of the law which has generated jurisprudence is that a butcher (retailer) was found responsible for selling meat infested with trichonellosis (horsemeat, CA Toulouse 112632, December 2000). It is noteworthy that the scandal of tainted blood in the 1990s, as well as the BSE crisis, has raised the issue of the responsibility of the government for not taking timely measures that would have prevented distribution of contaminated material.

One major question is the possible use of the French law 98-389 as an instrument for ensuring liability of producers and sellers of GMOs in the case of possible negative effects on consumers' health or the environment. While lawyers are still divided on that, it seems that GMO crops and genetically modified animals qualify as a "product" covered by the law, given the French interpretation of the EU directive, and the inclusion in the French law of agricultural goods. While this is still subject to controversy, jurisprudence suggests that a possible "defective" character of a GMO could also be taken into account, even though it has been approved by the various commissions involved in the EU and French process of approval. This would particularly be the case if labeling had been improperly done (labeling is codified by a French law that has translated the 1139/98 regulation). However, there are difficulties that raise questions on the practical role of this law. Possible accidental contamination of non-GM food by GMOs can occur. There is still disagreement on the definition of equivalent and non-equivalent products. The status of the development risk, which is a case of liability exclusion in the Directive, is still unclear in practice. The multiplicity of interventions in the food chain makes it difficult to assess who is responsible and which clauses of limitation of responsibility can be invoked. Perhaps more important, a liability action might be difficult for a victim of GMOs because of the burden of proof that is borne by the plaintiff. Scientific evidence might make it difficult to determine the cause of a possible damage. The ten-year limitation of the directive might also appear short in the case of GMOs. In brief, the legal issue of producers' liability in the case of GMOs, even when national laws provide more precision than the EU directive, is still very controversial and unclear. (Readers will find more details on this complex issue in Cassin (Cassin 1999)).

The directive aspires to protect victims in addition to promote improved food safety. It covers any defective product manufactured in or imported into the EU. The concept of strict liability (without fault) is introduced, the burden of proof is set on the injured party. However, the producer will not be liable if it is proven that at the time the product was placed on the market the defect which caused the damage did not exist, or if the state of scientific knowledge was not such as to enable the defect to be discovered. Since the 1999 amendment, producers of non-processed agricultural products are also liable without fault for damage to the health of individuals caused by defective products.

The BSE and dioxin crises have pressed the Commission to consider reforming some aspects of the Directive 85/374/EEC. This is the reason why the Commission has recently adopted a Green Paper on producer liability that reflects a consultation of all sectors. The aims of the paper are to collect information on the application of the Directive and to gauge reactions to possible revisions. The latter include aspects such as the goods and damage covered, the burden of proof, the existence of financial limits, the ten-year deadline for responsibility, suppliers' liability, and the lack of any obligation on producers to obtain insurance that characterizes the Directive.

Food safety from the economic side

Governments can use several policy instruments to protect consumers and alleviate market inefficiencies. Economists often distinguish mandatory regulations (e.g. minimum safety standards (MSS), including input standards, process standards or product-performance standards), information regulation (e.g. labeling) and liability enforcement. These different instruments are seen as means to circumvent some market inefficiencies such as informational inefficiencies (e.g. imperfect consumer information on the safety of products, or non-revelation of producer information) and insufficient safety efforts by producers.

Economic theory suggests that a system of market, regulatory and legal components can provide incentives to firms to produce safe food (Antle 1995). Economists often favor market-forces-based instruments. They rely on incentives for firms not to lose business reputation, market shares or sales revenues if consumers become concerned about safety problems with a firm's product.

When human health is at stake, public intervention often favors command and control instruments such as MSS. However, such instruments are sometimes costly and do not always pass cost–benefit analysis (Arrow et al. 1996). In some cases, instruments relying on consumers' information are preferable. When risk is deemed to be small and/or non-lethal, giving consumers the choice between different levels of risk at different prices may be economically efficient (Beales, Craswell and Salop 1981).

The economic theory develops some recommendations about liability, one of America's favorite instruments, that European regulators seem to be less fond of. Is there scope for changes in the EU food-safety regulation towards more responsibility-based instruments? In Europe, the Directive 85/374 is extended to defective agricultural products. However, only five countries among the fifteen Member States transposed the liability for defective agricultural products in their domestic law (see Annex 1 page 34 in the Green Paper (*Green paper on liability for defective products* 1999)). Very few Eastern-European countries looking towards membership in the EU transposed the legislation on this point.

Legal liability for damages generally has two goals: providing compensation to victims and limiting risks by creating incentives for lowering the probability and/or the severity of accidents. Two types of liability rules are usually distinguished by economists: the strict liability for compensating victims whatever the efforts made for limiting risks, and the negligence rule for punishing injurers only if the regulation was not respected (Shavell 1980). Liability appears to be a flexible tool, since each firm has some freedom for managing the risk. Under negligence rule, the regulator (and indirectly taxpayers) may be overexposed to liability if the regulation was deemed *ex post* insufficient or was not implemented. This is the issue that is raised by the many criticisms underlining the absence of regulatory reactions for the BSE outbreaks in Europe. (As it is impossible to identify the origin of the BSE contamination, the regulator in France appears to be the only "person" responsible for the victims' families (see Libération, "Inerties et entraves au Ministère de l'Agriculture", 26 Mars 2002). Pressure from lobbies for limiting stringent regulation in the nineties could suggest extension of liability to the animal feeding industry or to farmers' unions. For instance, the head of the French Agency against fraudulent products (DGCCRF) was regretting some resistances by the French Ministry of Agriculture for implementing policies: *"Nous nous sommes heurtés dans tous les cas à une profonde réticence du ministère de l'Agriculture. Ce dernier semble soucieux de privilégier des intérêts particuliers sans prendre suffisamment en compte la protection de la santé du consommateur"*, Comment for the Ministry of Finance, by Christian Babusiaux, DGCCRF, 24 Mai 1995.)

The threat of sanctions for regulator failures by independent Courts could force the regulator to put the regulation into effect. For instance, in 2001, the regulator's liability regarding water pollution was established by a Court due to its absence of real control regarding the pig facilities in French Brittany (Tribunal administratif de Rennes, 2 Mai 2001, n°97182).

Two conditions are necessary for reaching an efficient level of food safety via liability: (1) sufficient information for consumers for proving harm *ex post*, (2) sufficient funds from the injurer's side for a complete indemnification in case of a damage in order to avoid "judgment-proof" problems. Both "failures" are acute in the food system (Antle 2001).

Regarding imperfect information about harms, the origins of contamination with microbial contaminants or pesticide are often impossible to prove for isolated consumers. Indeed, the precise liability is difficult to establish in a vertical supply chain including numerous farmers, processors and retailers. Moreover, consumers eat many different types of products. Moreover, how to determine a product liability while the food has disappeared once it was eaten? Consumers may be careless regarding food freezing or cooking. Eventually, incubation is sometimes very long as for the Creutzfeldt-Jacob disease (BSE) or cancer linked to chemicals (dioxin). Consumers need to have credence in goods of which they cannot judge the safety at the time of purchasing. For all those goods, preserving brand reputation is not sufficient for entailing a firm's effort. For producers of processed food or for retailers, the preservation of a brand reputation may influence the level of prevention. All these reasons suggest that liability is limited due to imperfect information. However, the mandatory traceability or certification could be a way to develop an efficient liability process, since it becomes possible to detect fraudulent sellers or tainted food. For instance, traceability is already used for food recall systems (such as those implemented by the French Institut de Veille Sanitaire) in case of contamination incidents. As to liability, product recall may be costly to firms in terms of reputation (e.g. on firms' market valuation, see Salin and Hooker (Salin and Hooker 2001)).

A food recall is a voluntary action carried out by industry in cooperation with federal and state agencies. It may be initiated either by the firm or at the request of the regulating agencies. The purpose of a recall is to remove products from markets when there is evidence that the product is contaminated, adulterated or misbranded. To this regard, they constitute information on product harmfulness that could be used by plaintiffs.

The link between liability and traceability raises the issues of *(i)* the way to finance certification taking into account the liability (Crespi and Marette 2001), (*ii*) new contractual relationship between suppliers and clients including scientific tests and monitoring, (*iii*) new technologies and investments for traceability that could influence market concentration and/or vertical integration in the agro-food chain. Public labeling for GMOs may imply false messages and cheating on products, requiring sanctions (McCluskey, 2001). The labeling "GMO-free" seems very fragile according to the imperfect testing procedures.

With respect to imperfect compensation, legal liability faces the issue of the "judgment-proof" injurer, that is an injurer unable to pay some portion of the losses to victims. The fact that a potential liability of an injurer is bounded by its wealth and the limited liability, by reinforcing risk, has implications on the prevention activity (Shavell 1986). Many prevention efforts require better equipment. For instance, prevention efforts by industrial firms or farmers for improving safety or reducing pollution are very difficult to implement, even in developed countries, mainly because of the prohibitive costs. Firms invest in prevention if their expected profits are high enough. In the meat sector, Antle (Antle 2001) and MacDonald and Crutchfield (MacDonald and Crutchfield 1996) showed that small processing plants suffered from implementing safety regulations and/or HACCP plans due to a resulting high fixed cost. Prevention costs and the degree of competition among firms are also a key issue in determining producers' solvency. Box 4 shows that insolvency may be strategically chosen by firms according to the market structure (i.e., the number of firms).

Concentration in the industry influences the profit of firms, and indirectly the wealth of firms and their probability of being "judgment-proof". The agro-food sector is characterized by the coexistence of multinational companies wielding oligopolistic/oligopsonistic power and able to cover damages and farmers with a very limited ability for indemnifying third parties or consumers. The concentration ratio is high for many food industries in Europe and in the US (Cotterill 1999) allowing the use of liability. However, the limited farmers' size suggests that liability could be extended to their upstream suppliers (as the agrochemicals firms) or the downstream processors. On the other hand, liability as a regulatory instrument might play a significant role in the concentration of the industry. An interesting illustration, with transnational consequences, is the case of the company Aventis, which recently sold its share of Aventis CropScience to Bayer, partly due to pending liability payments because of the new StarLink maize, a genetically modified organism (GMO). Traces of the GMO were found in processed foods, the costs of which may force Aventis CropScience to compensate farmers and manufacturers up to US$ 200 million in the USA (see Taylor and Tick (Taylor and Tick 2001) for details). Aventis faced too many financial difficulties after the resulting withdrawal of the StarLink maize. This litigation in the USA had a major effect on the world market structure for the agrochemicals with the recent merger between two European firms, Aventis and Bayer. Despite the regulation harmonization among countries, the foreign regulation may have direct effects on a domestic market. Four recent mergers have led to a rapid increase in the three-firm ratio (CR3, namely the sum of the market share of the three largest firms) with a shift from 34% in 1998 to 60% in 2000 concerning the worldwide trade for agrochemicals. Among the main reasons for those recent mergers are: *(i)* new environmental standards and pesticide-residue regulation entailing high expenditures on R&D, *(ii)* increasing risk of liability suits due to food safety or the environment, *(iii)* consumer reluctance toward innovations such as GMOs, which deters some farmers from adopting these products. Harhoff, Régibeau and Rockett (Harhoff, Régibeau and Rockett 2001) underline the dilemma for this industry between a stringent regulatory approval process and the risk of increasing market concentration. This dilemma also exists for the liability tool due to the risk of insolvency that depends on the market concentration (see Box 4).

Box 4: Liability, product information, and solvability

Coestier, Gozlan and Marette (Coestier, Gozlan and Marette 2002a; Coestier, Gozlan and Marette 2002b) provide theoretical models designed for the analysis of limited liability. Under a judgment-proof scenario, victims are not fully compensated due to the limited wealth of a firm. The costly prevention activity may dramatically reduce the available funds for indemnification in case of a damage. Victims compensation can also be limited due to the small size of involved producers. The delegation of risky activities from big firms to small and medium firms with a pronounced limited liability weakens the effectiveness of any legal liability regime.

An analysis of the incentives of firms to invest in prevention under alternative public regulations is proposed, in the case of consumers unaware about the safety risk. An efficient regulation needs carefully consider the following aspects. First, the prevention activity affects the profit of firms: a higher effort reduces both the probability of an accident and the profits that are available to pay damages; thus, higher effort increases the probability of being judgment-proof. Second, the profits of firms are affected by the market structure: the more concentrated the market, the higher the profits and the higher the assets available for compensation; higher profits reduce the probability of being judgment-proof. A unified framework for studying the impact of legal liability on prevention is provided, making it possible to emphasize the strategic use of insolvency by firms.

It is shown that whatever the market structure, under alternative liability rules, the private optimal level of effort depends on the perspective of profits and more precisely on the maximum willingness to pay for consumers with respect to the magnitude of damages. For certain parameter values, incentives to invest in prevention are diluted: under strict liability where the injurer is liable regardless of his effort, firms either underinvest or overinvest in prevention. The overinvestment in prevention appears as a pernicious effect of liability under potential insolvency: it is the result of the strategic use of the limited wealth and limited liability by firms. It can be corrected with the negligence rule under which the injurer is liable only if the level of "due care" was not implemented (Coestier, Gozlan and Marette 2002a). When the magnitude of damages is large, the negligence rule as strict liability leads to underinvestment in prevention.

In a second paper, the strict liability is compared with a policy that provides information about the product risk to the consumers (Coestier, Gozlan and Marette 2002b). Providing information avoids the judgment-proof problem. Indeed, the risk is internalized in the demand by consumers. When the potential damage is too large, consumers prefer to avoid purchasing, while being "judgment-proof" would emerge under strict liability. However, it is shown that providing information can be dominated by alternative policies such as strict liability, as an over investment in safety results in a higher welfare than a market closure.

The empirical study by Buzby and Frenzen (Buzby and Frenzen 1999) underscores the very limited use of liability in the USA and UK for food contamination. In particular, they mention the ambiguity about the nature of microbial contamination ("natural and to be expected", or "adulterant that should be controlled"), which is a key issue in litigation, and previous cases at the federal level in the United States have not provided a "consistent interpretation of liability". For example a 1974 case ruled *Salmonella* in chicken as a natural contaminant, while a 1994 decision made *E. coli* O157:H7 in beef an adulterant, due to the fact that pathogens would be eliminated by "proper cooking" in the first situation, but not in the case of beef, which is often "lightly cooked". Second, legal costs, time and difficulties concerning the burden of proof discourage in practice most individual people to seek legal redress for food-borne illness. Legal incentives may work better in outbreak situations than in individual cases, since the "causality relation" is then easier to infer for health authorities. Third, the US legal system "provides lawyers with incentives to choose their cases selectively and to strive for high settlement awards", resulting in frivolous suits and excessive or fraudulent consumer claims. The limited use of liability for food contamination is confirmed for France by Collart Dutilleul (Collart Dutilleul 1997; Collart Dutilleul 1998).

Along with reasons (1) and (2), the existence of the public indemnification (as for instance for the "mad-cow" disease outbreaks in the UK and France) and the existence of social security tend toward a diminished role for tort liability (Viscusi 1989). The evaluation of the damage for some contaminations (like *Salmonella* or *Listeria*) may vary a lot. (For instance, some ERS estimations of the economic loss linked to *Salmonella* in the US in 1995 range from $0.9 billion to $12.2 billion.) At the farm level, some prevention efforts are sometimes difficult to implement due to the absence of knowledge, as for pesticide application or soil pollution.

Finding the right policy instrument is a titanic task as is illustrated in Box 5. Even though the case described deals with a very stylized case, in particular with a very simple competition structure, the optimal instrument between mandatory standards, labels and liability depends on several factors. It shows, in particular, that reputation-based instruments can be efficient if consumers find out rapidly that a product was unsafe after consumption. If a hazard appears in the long run, there is generally the need for more command and control policy instruments (Shavell 1987).

Box 5: Information and optimal regulation

One of the major problems that economists face when analysing the efficiency of the various forms of regulation in the product-safety area, is the structure of information. Optimal instruments are likely to be different if consumers are informed on the safety of the products (perfect information), if they can find out that a product is tainted after some research or investment in time (search good), if they can find out immediately after consumption (experience good), or if they never find out, or with a very long delay, or if they can never be sure that a particular good caused the disease (credence good). What makes the issue more complicated is that the optimal regulation for a given information structure can also depend on the structure of competition of the industry. Competitive industry will face price pressures that interfere with the reputation signals that they can deliver to consumers, which also has price effects (a quality signal can take the form of a low introductory price, of a high and non-imitable price, or of investment in publicity, etc.).

In order to illustrate the optimal regulation to provide food safety under various information structures, Marette, Bureau and Gozlan (Marette, Bureau and Gozlan 2000) have assumed that a firm faced no other price constraint than the willingness to pay of its customers (that is, it acts as a monopoly and sets its price unilaterally). In their model, trade takes place over two periods. A common discount factor $\delta \geq 0$ is used for valuing the second-period gains relative to the first-period ones. The marginal cost of production c is constant, regardless of the safety of the product. A product is either harmful or harmless, and a higher level of safety effort made by the seller increases the probability of it to be harmless. By selecting a level of effort $\lambda \in [0,1]$, the seller incurs a fixed cost equal to $f\,\lambda^2/2$ in the first period. This framework accounts for both a moral-hazard effect and an adverse selection effect when consumers are imperfectly informed. Indeed, the level of safety depends on the seller's effort, which refers to moral hazard. However, when $\lambda < 1$, the seller cannot totally control the final safety of the product, which refers to adverse selection. Consumers purchase either one or zero unit of the good. Acquiring harmful products results in a zero utility for consumers. Heterogeneous willingness to pay for safe goods is represented by a parameter $\theta \in [0,1]$ distributed over the population of consumers. In the first period, if prior to purchase a consumer detects that the good is harmful, he/she does not acquire any product, resulting in a zero utility. The consumer will acquire a good in the second period only if he/she acquired a harmless product in the first period.

Solving the optimal program of the producer leads to the conclusion that: *(i)* when the cost of the safety effort f is large the safety effort is systematically lower than the socially optimal level (under perfect information as well as with experience goods); *(ii)* despite experience by consumers, imperfect information is likely to lead to a lower safety effort than under perfect information; *(iii)* with experience goods, the seller chooses not to signal the safety of its product. Nevertheless, the prospect of sales in the second period is an incentive for providing a significant safety effort, while, with credence goods, market forces result in an absence of effort and a market failure.

Regarding the optimal policy instruments, they show that a minimum safety standard (MSS) can be a useful tool for correcting a sub-optimal level of safety under both perfect and imperfect information. With experience goods, an MSS leads to shifting from a separating to a pooling equilibrium. This results in a higher safety effort, but may increase the welfare losses due to the pricing strategy of the seller. With credence goods, an MSS is a necessary and efficient tool when the cost of the safety effort is low. However, for large values of f an MSS does not make it possible to avoid market closure.

A label is potentially a useful instrument for reducing market inefficiencies in the case of credence goods, provided that the regulator or a third party mandated by consumers, has the ability to verify the safety effort. This may require specific means for monitoring the production process.

Under *perfect information*, liability enforcement is irrelevant since consumers are aware of the characteristics of the products by assumption, and dangerous products are not purchased. Enforcement of liability policy in the case of *credence* goods runs into the lack of conclusive evidence that a particular disease results from a well-identified source. If the regulator verifies the production process, punitive damage for misrepresentation with a credence good has effects similar to those of an MSS.

In the case of *experience* goods, the regulator may implement a punitive damage so that it prevents the seller from selecting the pooling equilibrium with harmful products. The punitive damage must therefore be larger than the profit under pooling equilibrium without any effort. In such a case, liability enforcement may be a policy instrument that dominates the MSS in terms of welfare. Indeed, it makes it possible to benefit from the safety-revelation mechanism, while avoiding possible prohibitive compliance and enforcement cost of the MSS. The combination of liability enforcement and MSS may show some form of social optimality. The setting of an appropriate MSS could result in a socially optimal level of effort. Simultaneously, liability enforcement deters the sale of harmful products. This policy mix therefore leads to a market equilibrium, which is similar to the one under perfect information, even in the presence of experience goods.

Conclusion

Recent food crises, together with the development of new production techniques and the international pressures provided by the WTO Uruguay Round agreements dealing with product quality and safety, have urged the European Union to reform safety regulations. The White Paper is a fundamental step in that direction. It has led to the implementation of a centralized authority, and is likely to lead to future directives and national regulations.

It is noteworthy that command and control as well as informational instruments are more central in the White Paper, and more generally in the EU food-safety regulation, than incentive-based instruments, such as product-liability laws. The recent arbitration of the EU parliament for minimizing the liability of GMO producers suggests that the liability instrument is not as favored by regulators as it might be in other countries. Economists, however, tend to consider that if a firm that circulated products making people ill has to pay financial compensation as well as punitive damage, courts and legal fees, it will have more incentives to invest in food safety than if its managers are only liable in front of a penal court. However, in order to be efficient, product-liability laws must specify the exact circumstances under which firms are held liable. The present EU legislation on liability, and the ongoing thought that has led to a Green Paper on liability suggest that there are still many gray areas as to whether or not food safety issues could fall under liability laws (*Green paper on liability for defective products* 1999). The uncertainty surrounding the liability of producers in the case of a possible problem with GMOs that have been approved by the EU regulator illustrates the ambiguities of the present legislative framework in the case of a development risk. The development of traceability in the food supply chain may reinforce the effective role of liability.

Potential liability is part of the expected costs of the firm that would not take enough precautions regarding product safety. A firm will invest in safety up to the point where the marginal cost of safety equals the marginal expected benefits of reducing the risk of financial compensation and possible punitive damage. Therefore, the legal system could provide optimal deterrence if the firm correctly anticipates the compensation. Despite the limitations to the efficiency of liability as a regulatory instrument, we believe that there is a need for more investigation on the scope for which liability can be an efficient instrument, and that this issue is perhaps overlooked in the case of food safety. Instruments like incentive-based mechanisms have limitations, but they are also very powerful instruments that could minimize the administrative and regulatory burden. If liability is not always the right solution, the costs of command and control instruments for the society as a whole should not be underestimated.

In the various EU initiatives, government regulation is not the only approach deserving consideration, with measures ranging from voluntary practices to codes of good conduct, private standards, labeling and economic incentives. Still, the issues are complex and the required policy response unclear. The appropriate response is especially difficult to ascertain in cases where there are strong consumer concerns and insufficient or uncertain scientific evidence of health risk, as for GMOs.

References

Agriculture in the WTO: the role of product attributes in the agricultural negotiations, 2001. International Agricultural Trade Research Consortium, University of Minnesota, St. Paul, MA. IATRC Commissioned Papers no. 17.

Antle, J. M., 1995. *Choice and efficiency in food safety policy*. AEI Press, Washington, DC.

Antle, J. M., 2001. Economic analysis of food safety. *In:* Gardner, B. L. and Rausser, G. C. eds. *Handbook of agricultural economics, volume 1*. Elsevier, Amsterdam. Handbooks in Economics no. 18.
Arrow, K. J., Cropper, M. L., Eads, G. C., et al., 1996. Is there a role for benefit-cost analysis in environmental, health and safety regulation? *Science,* 272, 221-222.
Beales, H., Craswell, R. and Salop, S., 1981. The efficient regulation of consumer information. *Journal of Law and Economics,* 24, 491-544.
Bureau, J. C., Gozlan, E. and Marette, S., 1999. *Food safety and quality issues: trade considerations*. Organisation for Economic Co-operation and Development, Paris.
Buzby, J. C. and Frenzen, P. D., 1999. Food safety and product liability. *Food Policy,* 24 (6), 637-651.
Cassin, I., 1999. Les organismes génétiquement modifiés et le nouveau régime de la responsabilité du fait de produits défécteux. *Gazette du Palais,* 22-23 (1), 99.
Coestier, B., Gozlan, E. and Marette, S., 2002a. Prevention limited liability and market structure. *In: Papers of the 5th INRA-IDEI conference on: industrial organization and the food processing industry, Toulouse, France, June, 14-15, 2002*. UMR Economie Publique, INRA, Paris.
Coestier, B., Gozlan, E. and Marette, S., 2002b. Product safety: liability rule versus information regulation. *In: Papers of the 5th INRA-IDEI conference on: industrial organization and the food processing industry, Toulouse, France, June, 14-15, 2002*. UMR Economie Publique, INRA, Paris.
Collart Dutilleul, F., 1997. Regards sur les actions en responsabilité civile à la lumière de l'affaire de la vache folle. *Revue de Droit Rural,* 252, 227-233.
Collart Dutilleul, F., 1998. Les analyses en agroalimentaire et le droit de la responsabilité civile. *Revue de Droit Rural,* 266, 450-455.
Cotterill, R. W., 1999. *Continuing concentration in food industries globally : strategic challenges to an unstable status quo*. Dept. of Agricultural and Resource Economics, University of Connecticut, Storrs. Research Report University of Connecticut, Food Marketing Policy Center no. 49.
Crespi, J. M. and Marette, S., 2001. How should food safety certification be financed. *American Journal of Agricultural Economics,* 83 (4), 852-861.
Green paper on liability for defective products, 1999. Available: [http://europa.eu.int/comm/internal_market/en/update/consumer/greenen.pdf] (6 Mar 2003).
Harhoff, D., Régibeau, P. and Rockett, K., 2001. Some simple economics of GM food. *Economic Journal,* 111, 265-291.
MacDonald, J. M. and Crutchfield, S., 1996. Modeling the costs of food safety regulation. *American Journal of Agricultural Economics,* 78 (5), 1285-1290.
Marette, S., Bureau, J. C. and Gozlan, E., 2000. Product safety provision and consumers' information. *Australian Economic Papers,* 39 (4), 426-441.
Salin, V. and Hooker, N. H., 2001. Stock market reaction to food recalls. *Review of Agricultural Economics,* 23 (1), 33-46.
Shavell, S., 1980. Strict liability versus negligence. *Journal of Legal Studies,* 9, 1-25.
Shavell, S., 1986. The judgment proof problem. *International Review of Law and Economics,* 6, 45-58.
Shavell, S., 1987. *Economic analysis of accident law*. Harvard University Press, London.

Sheldon, I. M., 2002. Regulation of biotechnology: will we ever 'freely' trade GMOs? *European Review of Agricultural Economics,* 29 (1), 155-176.

Taylor, R. and Tick, J., 2001. *The StarLink Case: issues for the future*. Resources for the Future, Washington, DC.

Viscusi, W., 1989. Toward a diminished role for tort liability: social insurance, government regulation, and contemporary risks to health and safety. *Yale Journal on Regulation,* 6, 65-107.

White paper on food safety, 2000. Available: [http://europa.eu.int/comm/dgs/health_consumer/library/pub/pub06_en.pdf] (6 Mar 2003).

10

A review of empirical studies of the trade and economic effects of food-safety regulations

Norbert Wilson[*]

Introduction

This paper is a synthesis of the empirical work analyzing the trade (e.g. trade volume) and economic (e.g. welfare) effects of food-safety regulations. Several papers provide a descriptive discussion of the general issues of food-safety and international trade (e.g. Henson et al. (Henson et al. 2000), Hooker and Caswell (Hooker and Caswell 1999), IATRC (*Agriculture in the WTO: the role of product attributes in the agricultural negotiations* 2001), OECD (*Food safety and quality issues: trade considerations* 1999), Roberts, Josling and Orden (Roberts, Josling and Orden 1999), Thilmany and Barrett (Thilmany and Barrett 1997)). While these studies provide interesting frameworks and conceptualizations of the trade and economic effects of food-safety regulations, the papers fail to provide quantitative evidence of the effects of regulations.

This paper is a review of the few empirical papers on the food-safety regulations. The emphasis of the empirical literature is to begin toestablish a benchmark of the effects of the food-safety regulations. This literature review provides the results of analyses of a limited group of food-safety regulations.

The regulations restricting trade to maintain food safety, in addition to regulations affecting the trade of products that may carry pests or disease that harm plant or animal life and health (and loosely, the environment), fall under the rubric of the Agreement on the Application of Sanitary and Phytosanitary Measures (the SPS Agreement) of the World Trade Organization (WTO). In the broadest sense, food-safety regulations affecting international trade are non-tariff barriers (NTB). OECD (*Measurement of sanitary, phytosanitary and technical barriers to trade* 2001) drew attention to NTB and suggested four reasons for empirical research on NTB:

- Domestic regulations may constitute major trade impediments and their use is proliferating. However, these NTBs may simply become more visible because of international scrutiny or more trade-restrictive because of the decrease in tariffs.
- Quantification of the economic effects of SPS and technical regulations is an important step in the regulatory reform process (Regulatory reform in the agri-food sector 1997). Quantitative analyses help inform governments as to the cost of their SPS policies and provide the elements for defining more efficient regulations (Antle 1995).
- More satisfactory techniques for estimating the damage caused to a country by foreign regulations may help to solve disputes and may serve as a basis for calculating compensation claims.

[*] *OECD Directorate for Food, Agriculture and Fisheries, 2 Rue André Pascal, 75775 Paris Cedex 16, France*

A.G. J. Velthuis et al. (eds), New Approaches to Food-Safety Economics, 107-115.

- Sectoral studies suggest that technical regulations in developed countries constitute a considerable obstacle to agricultural food and feed exports of developing countries (Cato and Lima dos Santos 1998; Otsuki, Wilson and Sewadeh 2001a).

Despite the relevance of understanding the trade and economic effects, the literature on the trade and economic effects of SPS regulations, particularly food-safety regulations, is small. Thilmany and Barrett (Thilmany and Barrett 1997) stated, "Currently, little is understood about how regulatory barriers affect trade and investment volumes, nor how they affect the economic welfare of various global consumer populations." Since the mid-1990s, researchers have generated additional studies of the trade and economic effects of regulations of the SPS Agreement. At most this literature just begins to establish a benchmark of the effects.

Study 1

Otsuki, T. J, S. Wilson, and M. Sewadeh (Otsuki, Wilson and Sewadeh 2001b). "What Price Precaution? European Harmonisation of Aflatoxin Regulations and African Groundnut Exports"

What was the question?

The authors questioned the trade effects on nine African countries (Chad, Egypt, The Gambia, Mali, Nigeria, Sudan, Senegal, South Africa, and Zimbabwe) of a proposed, more stringent food-safety regulation in the EU (14 EU Member States except Greece). The proposed regulations would harmonize all Member States of the EU to a regulation of 2 ppb for aflatoxin B1, a carcinogen found in groundnuts. The proposed change in regulation would have been more stringent for all but four of the Member States.

What was the method employed?

The authors used the gravity model to estimate bilateral trade flows and the effect of regulations on these flows. The model "specifies that a flow from origin [*j*] to destination [*i*] can be explained by economic forces at the flow's origin, economic forces at the flow's destination, and economic forces either aiding or resisting the flow's movement from origin to destination" (Bergstrand 1985). A simplified specification of the model as presented in Otsuki, Wilson and Sewadeh (Otsuki, Wilson and Sewadeh 2001b) is[i]:

$$\ln(M_{ijkt}) = b_{ok} + b_{1k}\ln(GNPPC_{it}) + b_{2k}\ln(GNPPC_{jt}) + b_{3k}\ln(DIST_{ij}) + b_{4k}\ln(ST_{ikt}) + b_{5k}\ln(RAIN_{jt}) + b_{6k}COL_{ij} + b_{7k}YEAR + \varepsilon_{ijkt}.$$

M_{ijkt} was the trade flow in the amount of product k to EU Member State i from African country j in year t[ii]. The products were edible groundnuts, oil, and oilseed. The b parameters were the coefficients to be estimated, while the error term ε_{ijkt} was assumed to have a zero mean and to be normally and independently distributed. $GNPPC_{it}$ and $GNPPC_{jt}$ were the real per capita gross national products (GNP) in EU Member Country i and African country j in year t adjusted to the 1995 US dollar. $DIST$ was the geographical distance between countries i and j. ST_{ikt} was the maximum aflatoxin (Aflatoxin B1) level imposed on groundnut products by country i for product k for year t (since data were only available for 1995, the authors used that level for all years, assuming no change in maximum levels). $RAIN_{jt}$ was the average rainfall in African country *j*. The authors included the average rainfall because moisture levels positively influence aflatoxin levels during storage. COL_{ij} was a dummy variable for a colonial tie between countries *i* and *j*. *YEAR* was a linear time trend (with 1989=1 to 1998=10) to account for technological change.

The regression was a pooled regression with dummy variables for oilseeds and oil interacted with the coefficients of per capita GNP, rainfall, regulations and the intercept. The model also had a fixed effects structure where the groups were defined by the exporting country.

What were the results and implications?

The results showed that the per capita GNP of the EU countries had a positive and significant effect on exports for all products. The variable representing colonial ties was significant and positive. The variable representing the regulation was significant and positive for edible groundnuts and oil. The authors also estimated the model as unpooled five-year, rolling-average blocks, which generated elasticities of increasing value, suggesting that the regulation became more substantial over time. Using the estimated elasticities (covering the entire period) of the regulations and the trade volume and prices of 1998, the authors showed that making the regulation more stringent at 1 ppb, the estimated loss of value for African exports would be 482,400 US$ or 72 percent of the 1998 value. If the EU countries adjusted their regulations to the proposed EU regulation of 2 ppb, then the loss to African exports would be 238,900 US$ or 36 percent of the 1998 value. If all of the Member States of the EU adjusted their regulations to 9 ppb, the international standard that Codex Alimentarius[iii] suggested, the increase in trade value would be 480,600 US$ or 72 percent of the 1998 value.

Study 2

Otsuki, T., J. S. Wilson, and M. Sewadeh (Otsuki, Wilson and Sewadeh 2001a), "Saving Two in a Billion: Quantifying the Trade Effect of European Food Safety Standards on African Exports."

What was the question?

Otuski, Wilson, and Sewadeh investigated the effect on the value of trade flows of a proposed, harmonized regulation on maximum allowable aflatoxin levels for two food product groups: i) cereals and cereal preparations and ii) dried fruits, nuts, and vegetables. Specifically the authors made comparisons among the status-quo regulation; the proposed EU regulation, which would harmonize the Member States and was more stringent than the status quo for most Member States; and the suggested Codex regulation, which was less stringent for most Member States. The results of the changes in the regulation were linked to the differences of estimated health outcomes in terms of number of liver cancer deaths resulting from the different maximum aflatoxin levels. The countries included in the model were 15 European countries (Norway and the Member States of the EU except Greece) and nine African countries (Chad, Egypt, the Gambia, Mali, Nigeria, Senegal, South Africa, Sudan, and Zimbabwe).

What was the method employed?

The authors used a similar gravity model as presented in paper 1 to look at the effect of aflatoxin regulations on the value of trade flows for two food product groups.

$$\ln(V_{ijkt}) = b_{ok} + b_{1k}\ln(GNPPC_{it}) + b_{2k}\ln(GNPPC_{jt}) + b_{3k}\ln(DIST_{ij}) \\ + b_{4k}(COL_{ij}) + b_{5k}YEAR + b_{6k}\ln(ST_{ikt}) + \varepsilon_{ijkt}.$$

The variables in this model were described in the methods section of paper 1, except for the dependent variable. The dependent variable V_{ijkt} was the value, not the volume, of trade of product k imported by country i and exported country j in year t.

The authors stated "Dummy variables for exporting countries are included in the model in order to control unobserved factors such as environment and product quantity, which may vary across these countries" (p. 505). Therefore, additional dummy variables were included in the model. The specification for the gravity model also included a fixed-effect specification for the importing countries.

What were the results and implications?

The models were estimated for i) cereals and cereal preparations, and ii) dried fruits, nuts, and vegetables, separately. For the cereal and cereal-preparation model, the per capita GNP of European countries, the regulation and the colonial-tie dummy variables were positive and significant at the five-percent level. The distance variable was negative and significant at the five-percent level. For the dried fruits, nuts and vegetable model, per capita GNP for European countries, the aflatoxin regulation, and the colonial tie were all positive and significant at the five-percent level. The per capita GNP for the African countries variable was also positive but significant at the ten-percent level. For the dried fruits, nuts and vegetable model, the distance variable was negative and significant at the five-percent level.

These coefficients showed that an increase in the per capita GNP would increase the import of the products. The positive sign on the regulation suggested that increased stringency of the regulation, that is lowering the maximum allowable level of aflatoxin B1, would lower the trade of the products. The colonial tie had a positive effect on the trade of goods between countries. The negative sign on the distance suggested that countries that were more distant trading partners traded less than trading partners closer together.

The authors separated the data into three groups: i) coconuts, Brazil and cashew nuts; ii) groundnuts and other edible nuts; and iii) dried and preserved fruit. Two of the estimated elasticities for the regulation (because of the double-log specification, the elasticity is the estimated coefficient from the regulation variable) were positive and statistically significant at five-percent significance level - (groundnuts and other edible nuts; and dried and preserved fruit) and at the ten-percent level (coconuts, Brazil and cashew nuts). The elasticity of the regulation for dried and preserved fruit was not significant. The elasticity of the regulation on the trade of groundnuts and other edible nuts was larger than the elasticity for Brazil and cashew nuts. Given that most of the concern for aflatoxins has to do with groundnuts, the result was not surprising.

Given the estimated elasticities the authors calculated the impact of harmonizing the various EU regulations to the proposed, more stringent policy or to the less stringent Codex standard. In 1998 the value of cereal and cereal-product exports from Africa to Europe was 298 million US$. A move to the more stringent proposed EU regulation would generate a loss of 177 million US$ or 59 percent of the 1998 value. A move to the less stringent Codex standard would generate a gain in trade value of 202 million US$ or 68 percent of the 1998 value. The reduction in value of using the proposed EU harmonized regulation rather than the Codex standard would be 76 percent of the 1998 value.

The results were similar for the value of edible nuts. In 1998 the value of edible nuts from Africa to Europe was 472 million US$. A move to the more stringent, proposed EU regulation would generate a loss of 220 million US$ or 47 percent of the 1998 value. A move to the less stringent Codex standard would generate a gain in trade value of 66 million US$ or 14 percent of the 1998 value. The reduction in value when using the EU harmonized regulation rather than the Codex standard would be 53 percent of the 1998 value.

According to the estimated results and estimates from the Food and Agriculture Organisation (FAO), the number of lives saved from liver cancer from the more stringent aflatoxin regulations would be 0.9 lives saved per one billion persons. The loss in value of African food exports to the EU of moving to the more stringent, proposed EU aflatoxin regulation would be 340 million US$. Comparing the EU regulation with the Codex standard, the loss in value of African food exports would be 670 million US$ and the gain would be 2.3 lives saved per one billion persons.

Study 3

Wilson, J. S. and T. Otsuki (Wilson and Otsuki 2001), "Global Trade and Food Safety: Winners and Losers in a Fragmented System."

What was the question?

The question that Wilson and Otsuki attempted to answer was: what would be the effect of harmonizing of aflatoxin regulations on trade for cereal, edible nuts, and dried fruit on trade of these products? The authors also investigated the effects on the different importers and exporters.

What was the method employed?

The authors used a modification of the gravity model as given in papers 1 and 2 to look at the trade between 15 importing (4 developing) countries and 31 exporting (21 developing) countries.

$$\begin{aligned}\ln(V_{ijt}) = b_o + b_1\ln(GNPPC_{it}) + b_2\ln(GNPPC_{jt}) + b_3\ln(DIST_{ij}) + b_4(ST_{it}) \\ + b_5(COL_{ij}) + b_6 EU_{ij} + b_7 ASEAN_{ij} + b_8 NAFTA_{ij} + b_9 MERCOSUR_{ij} \\ + b_{10} YEAR96 + b_{11} YEAR97 + b_{12} YEAR98 + \varepsilon_{ijt}.\end{aligned}$$

The descriptions of most terms in the model are given in the methods section of paper 1. EU_{ij}, $ASEAN_{ij}$ (Association of South East Asian Nations), $NAFTA_{ij}$ (North American Free Trade Agreement), and $MERCOSUR_{ij}$ (Southern Common Market) were dummy variables, which are equal to one if the exporting and importing countries were members of the trade union and equal to zero otherwise. The variables *YEAR96*, *YEAR97*, and *YEAR98* were dummy variables representing the different years in the study. The authors limited the data set to the years 1995 to 1998, which was a shorter timeframe relative to the timeframe of the other two papers.

What were the results and implications?

The model was used for i) cereals, ii) edible nuts, and iii) dried and preserved fruit, separately. For the cereals model, the coefficients of per capita GNP of the importer, the regulation, the dummy variables for colonial ties, EU membership, and MERCOSUR membership were positive and significant at the one-percent significance level. The coefficient of the per capita GNP of the exporting country was also positive but significant at the five-percent level. The coefficients of the distance variable and the dummy variable for membership in NAFTA were negative and significant at the one-percent level.

For the edible nuts model, the coefficients of the variables per capita GNP for importing nations, the regulation, and the dummy variables for colonial ties and EU membership were all positive and significant at the one-percent level. The coefficient of the distance variable was negative and significant at the one-percent level.

For the dried and preserved fruits model, the coefficients of per capita GNP of the importer and exporter, the dummy variables for colonial ties, EU membership, and MERCOSUR membership were positive and significant at the one-percent level. The coefficient for the dummy variable of membership in ASEAN and NAFTA were positive and significant at the five- and ten-percent levels, respectively. The coefficients of the distance variable and the time dummy for 1998 were negative and significant at the one- and ten-percent levels, respectively. However, the coefficient for the regulation was not statistically significant; thus, further analysis of the variable was omitted.

The authors provided scenarios in which they compared different settings of the aflatoxin regulation. Under the base scenario all importers had different aflatoxin regulations. Of the 15 importers only four had a regulation that was less stringent than the Codex standard. Of the six EU Member States, only two had regulations, which were more restrictive than the proposed EU regulation. The different cases were 1) all nations move to the proposed, more stringent EU regulation, 2) only the EU nations move to the proposed EU regulation, and 3) all nations move to the Codex standard. The authors found that if all importing nations would adopt the Codex standard the trade of cereal and nuts would increase by 6.1 billion US$ or 51 percent compared to the 1998 level. The result was 7.1 billion US$ (65 percent) more than the value in the case where only the Member States of the EU adopt the proposed EU standard. In the case where all importing countries adopted the proposed EU regulation, the trade under the Codex standard would be 12.2 billion US$ (or 67 percent) more than the 1998 level.

Study 4

Overton, B. J. Begin, and W. Foster (Overton, Beghin and Foster 1995). "Phytosanitary Regulation and Agricultural Flows: Tobacco Inputs and Cigarettes Outputs"

What was the question?

Overton, Begin and Foster stated that at the time of publishing this paper, Germany, Italy, and Spain restricted the level of maleic hydrazide to 80 ppm in domestic and imported cigarettes[iv]. Maleic hydrazide is a growth inhibitor used in the production of tobacco. However, the regulation restricted the presence of the chemical in the final product, not on unprocessed tobacco. Because different levels of the market are affected by the regulation, the authors investigated the trade and economic effects of the maximum residue levels on the input and output levels. In particular, the authors simulated the effects of making more stringent EU regulation on tobacco-growing and tobacco-manufacturing industries on production costs, factor demands, and trade flows.

What was the method employed?

The authors used a partial equilibrium model to simulate the effect of a 10-percent reduction in the maximum residue level of maleic hydrazide in cigarettes. The model had supply and demand equations for US and EU cigarettes, assuming a constant elasticity of substitution for inputs and constant returns to scale. The model had derived demand equations for tobacco (from both the US, which contained the residue, and the EU, which did not have the residue) and other inputs. The authors made the level of the residue endogenous, which had to remain below the exogenous regulated level. The level of the residue was a function of the amount of US tobacco used and the quota lease rate US producers received, which was endogenous and a function of the US price and marginal cost.

What were the results and implications?

The authors estimated the impact of a 10-percent reduction in the maximum residue level under two scenarios:

- The US government maintained the pre-policy price of US tobacco by lowering the tobacco quota, which would cause the US marginal cost to increase, and
- the US tobacco price was allowed to fall by holding the amount of quota constant.

Under the first scenario, the authors showed that the demand for US cigarettes would fall by 0.085 percent while the demand for EU cigarettes would increase by 0.02 percent. Because of the constant returns to scale assumption, US tobacco-leaf exports increased compared to the EU. This result was because of the increase in EU cigarette production, which mitigated the overall loss of US tobacco (leaf and cigarettes) to the EU of 1.6 percent. US production would decline by 7.1 percent, and non-US tobacco production increased by 12.9 percent. The lower production of US tobacco would lead to a lower residue level of 2.8 percent.

In the second scenario, the quota remained fixed after the 10-percent reduction of the maximum residue level. The demand for US cigarettes would fall by 0.013 percent under the new price of US tobacco leaf. The price of US tobacco leaf would fall by 0.26 percent; thus, the EU and rest of the world demand for US tobacco leaf would increase by 0.86 and 0.47 percent. Despite the increase in the demand for US tobacco leaf, total exports of tobacco exports (leaf and in cigarettes) would decline by 1.51 percent. The increase in US tobacco leaf in EU cigarettes was the result of a substitution away from EU tobacco, a reduction of EU tobacco by 0.059 percent[v]. The residue level does not fall. This result showed that EU tobacco producers would be hurt, if only slightly, by the more stringent, maximum residue policy.

Assessment

The trio of gravity model papers provided some evidence that a food-safety regulation may have an impact on international trade. The results collectively provide evidence that a less stringent regulation may increase trade flows and consequently increase the income of foreign producers. The latter point is particularly important for developing countries. The effect of a more stringent regulation in some countries may reduce trade. However, the importance of that loss is unclear because of the mitigating effects of trade diversion. That is, while the trade to some countries is limited by a more stringent regulation, an importer with a less stringent regulation may purchase the diverted good. The diverted good may receive a lower price, but the trade is not lost necessarily as the gravity models predicted.

In general, the gravity model is not linked to the supply and demand of the product under study. The changes along or movements of the supply and/or demand curves to generate the changes in value (or volume) of trade that occur are not clear. Therefore, the welfare effects of the change in regulations are not known. The authors stated that they could not estimate the welfare effects of the different regulations with the gravity model. Nevertheless, estimating the welfare effects of changing regulations is important for understanding the differential effects of the regulations on different economic actors. Therefore, more work investigating the welfare effects of regulations is an important study area.

The four papers here and most of the other literature of NTB began with the hypothesis that the SPS regulations restrict trade. Overton, Beghin, and Foster (Overton, Beghin and Foster 1995) showed the surprising result that domestic producers might even be hurt by the regulation. Yet, the literature failed to explore the possibility that these regulations might improve trade. The trade-enhancing features of these regulations occur when measures reflect a movement toward greater harmonization or improved transparency. Disentangling the trade-enhancing effects of food-safety regulations, and more generally SPS regulations, from other factors that may enhance trade is difficult. Therefore, a useful area of research would be to look at whether harmonization and transparency can actually improve trade and welfare.

Another consideration for future research is an expansion of the geographic reach of the research. Studies that look at the effects of regulations from Northern countries on Southern countries would be beneficial. As seen in the SPS Committee, more and more regulations are being developed in Southern countries that are having an effect on other Southern countries.

Conclusion

Only a limited amount of research on the trade and economic effects of food-safety regulations exists. Further work needs to be done in this area. However, further research is hampered by the limited ability to generalize the various types of regulations. Another difficulty with research in this area is how to incorporate information from risk assessments into economic models appropriately. Such efforts require collaboration of food-safety scientists, regulators, and economists. This work will be beneficial in helping policymakers to develop food-safety policies that provide a nationally acceptable level of food safety with the least trade distorting policies.

References

Agriculture in the WTO: the role of product attributes in the agricultural negotiations, 2001. International Agricultural Trade Research Consortium, University of Minnesota, St. Paul, MA. IATRC Commissioned Papers no. 17.

Antle, J. M., 1995. *Choice and efficiency in food safety policy*. AEI Press, Washington, DC.

Bergstrand, J. H., 1985. The gravity equation in international trade: some microeconomic foundations and empirical evidence. *Review of Economics and Statistics,* 67 (3), 474-481.
Cato, J. C. and Lima dos Santos, C. A., 1998. European Union 1997 seafood safety ban: the economic consequences on Bangladesh shrimp processing. *Marine Resources Economics,* 13, 215-227.
Henson, S., Loader, R., Swinbank, A., et al., 2000. *Impact of sanitary and phytosanitary standards on developing countries*. Centre for Food Economics Research, University of Reading, Reading, England.
Hooker, N. H. and Caswell, J. A., 1999. A framework for evaluating non-tariff barriers to trade related to sanitary and phytosanitary regulation. *Journal of Agricultural Economics,* 50 (2), 234-246.
Measurement of sanitary, phytosanitary and technical barriers to trade, 2001. Organisation for Economic Co-operation and Development (OECD), Paris.
Otsuki, T., Wilson, J. S. and Sewadeh, M., 2001a. Saving two in a billion: quantifying the trade effect of European food safety standards on African exports. *Food Policy,* 26 (5), 495-514.
Otsuki, T., Wilson, J. S. and Sewadeh, M., 2001b. What price precaution? European harmonisation of aflatoxin regulations and African groundnut exports. *European Review of Agricultural Economics,* 28 (3), 263-283.
Overton, B., Beghin, J. and Foster, W., 1995. Phytosanitary regulation for US and agricultural trade flows: tobacco inputs and cigarettes outputs. *Agricultural and Resource Economic Review,* 24 (2), 221-231.
Regulatory reform in the agri-food sector, 1997. *In: The OECD report on regulatory reform. Volume I. Sectoral studies*. Organisation for Economic Cooperation and Development, Paris, 233-274.
Roberts, D., Josling, T. and Orden, D., 1999. *A framework for analyzing technical trade barriers in agricultural markets*. U.S. Dept. of Agriculture, ERS, Washington, DC. Technical Bulletin no. 1876.
Thilmany, D. D. and Barrett, C. B., 1997. Regulatory barriers in an integrating world food market. *Review of Agricultural Economics,* 19 (1), 91-107.
Wilson, J. S. and Otsuki, T., 2001. *Global trade and food safety: winners and losers in a fragmented system*. Development Research Group, The World Bank, Washington, DC.

---

[i] All of the model specifications presented in this paper are presented as they were presented in the original papers with only minor adjustments for clarity.
[ii] The natural log was taken for all variables, as indicated by the function ln(*) except for the dummy variables.
[iii] Codex, along with the International Office of Epizootics (OIE) and the International Plant Protection Convention (IPPC), provides standards, which are not legally binding as compared to regulations.
[iv] While a study of maximum residue levels in cigarettes does not qualify as a food-safety concern as interpreted from the SPS Agreement, the issue is closely related to food-safety concerns.
[v] Note the residue levels of EU-produced cigarettes were sufficiently below the maximum that an increase in US tobacco would not put EU cigarettes in danger of surpassing the maximum residue level.

11

International trade transparency: the issue in the World Trade Organization

João Magalhães[*]

Introduction

I was asked to participate in the discussion on international trade transparency with relation to food safety. Transparency provides essential information to consumers, producers and businesses. It also allows government authorities and private-sector economic actors to identify and deal with potential market-access problems.

In the World Trade Organization (WTO), food safety is especially addressed by the Agreement on the Application of Sanitary and Phytosanitary Measures (the SPS Agreement). One of the key goals of the SPS Agreement is to increase the transparency of sanitary and phytosanitary measures. Governments are required to notify other countries of any new or changed SPS measure, which has a "significant effect on trade". They have to set up "Enquiry Points" to respond to requests for more information on new or existing measures. Finally, governments also have to designate a central government authority (the National Notification Authority) as responsible for the implementation of the transparency requirements of the Agreement. Such increased transparency protects the interests of consumers, as well as trading partners, from hidden protectionism through unnecessary technical requirements.

The transparency obligations of the Agreement are contained in Article 5.8, Article 7 and Annex B. In addition, the SPS Committee has elaborated, and indeed recently reviewed, recommended procedures for implementing the transparency obligations of the SPS Agreement. These procedures clarify some of the Agreement's language and give guidance on how to notify (including how to fill in the notification formats), how to handle comments on notifications, and how to provide documents related to a notification. It also gives some guidance on the operation of National Notification Authorities and Enquiry Points, and on publication of regulations.

To help Members, in particular the developing and least-developed countries, to implement the transparency obligations, the Secretariat has put together a handbook titled: "How to Apply the Transparency Provisions of the SPS Agreement". It includes detailed descriptions of how to set up and operate Enquiry Points and National Notification Authorities, how to notify, models for letters of response, etc. The handbook has been revised in the light of the recent review of the recommended procedures.

The following paragraphs give a brief overview of the obligations contained in the Agreement. The handbook is available from the WTO web site and gives more detailed hands-on instructions and explanations on how to apply them.

---

[*] *World Trade Organisation, 154 Rue de la Lausanne, CH-1211 Geneve 21, Switzerland*

A.G. J. Velthuis et al. (eds), New Approaches to Food-Safety Economics, 117-124.

Publication of measures

All SPS measures that have been adopted have to be published promptly so that interested Members can become acquainted with them. Except in urgent situations, Members have to allow a reasonable period of time between the publication of a measure and its entry into force. This is to allow exporters, particularly developing countries, to adapt their products and methods of production to the new requirements.

Notification of measures

Members have the obligation to make SPS measures known if they:

- are new or changed; and
- are not based on an existing international standard, or a relevant international standard does not exist; and
- have a significant effect on trade.

This requirement covers measures that restrict trade as well as trade-facilitating measures. A notification should be made as soon as a complete draft of a proposed regulation is available, and when changes can still be made to take into account any comments received. For the sake of increased transparency, many Members even make measures known that are based on an international standard, or measures where it is not clear if they will have an impact on trade. Normally, regulations should be made known well before they enter into force. However, in urgent situations this may not be possible. Measures taken in urgent situations should nonetheless be made known immediately. In Figures 1, 2 and 3 the number of notifications per year and per country is given.

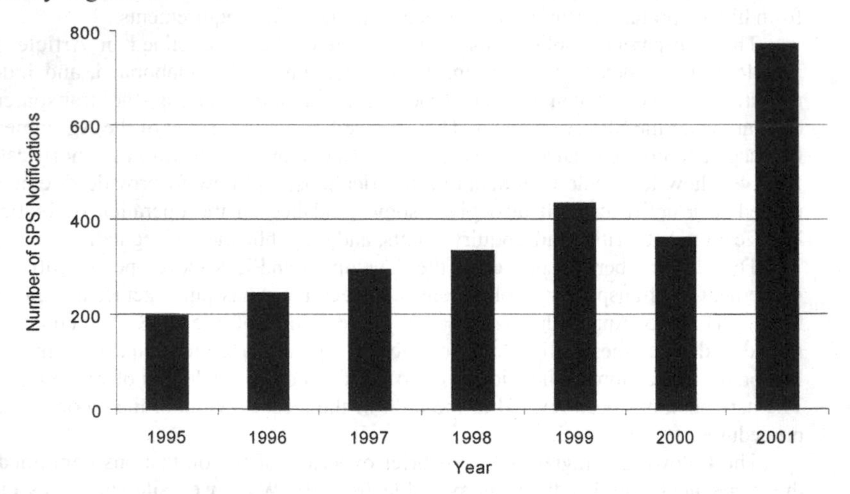

Figure 1. Number of SPS Notifications 1995-2001

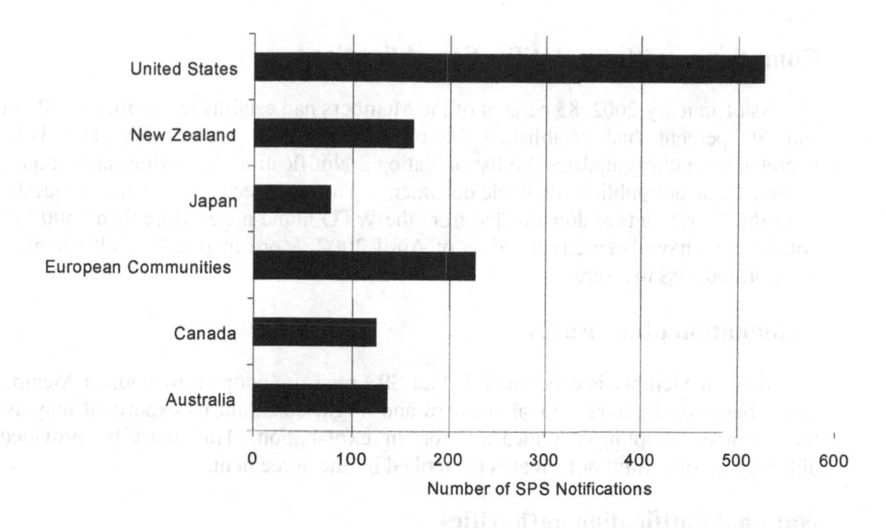

Figure 2. Number of SPS Notifications per selected developed country 1995-2001. The large number of recent US (pesticide-related) notifications is a result of EPA's implementation of changes in US pesticide laws enacted in 1996 and the development of new, lower-risk pesticides.

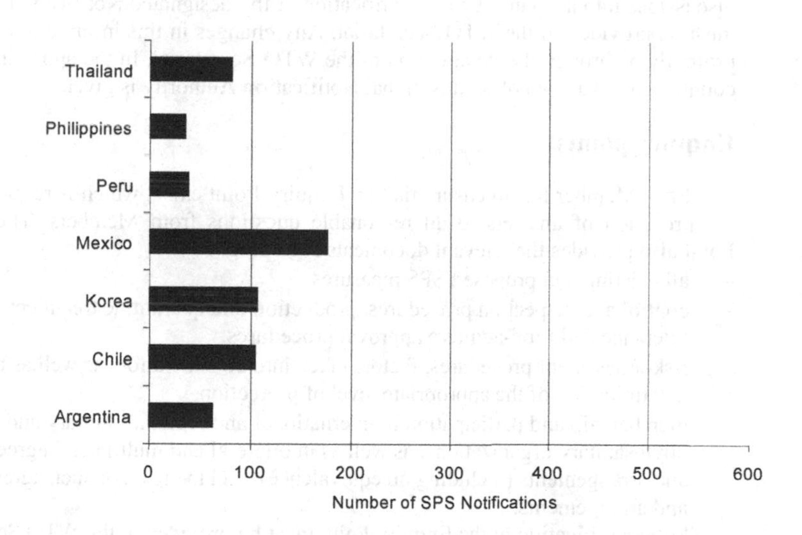

Figure 3. Number of SPS Notifications per selected developing country 1995-2001

Compliance with the notification obligations

As of January 2002, 85 percent of the Members had established an Enquiry Point, and 80 percent had established National Notification Authorities. The WTO Secretariat regularly updates the list of National Notification Authorities and Enquiry Points. These are publicly available documents. The most recent lists can be requested from the Secretariat, or downloaded from the WTO home page. More than 2800 SPS notifications have been circulated as of April 2002. More than half of all Members had notified SPS measures.

Explanation of measures

When a Member is concerned that an SPS measure adopted by another Member is not based on an international standard and might constrain its exports, it may ask the Member adopting the measure for an explanation. This must be provided, although no time limit or format is prescribed by the agreement.

National notification authorities

Members have to designate a single central-government authority responsible for the implementation of the notification obligations. This includes making draft measures known, providing copies of the proposed regulations, and receiving comments, discussing them upon request, and taking the comments and the results of discussions into account. A full identification of the designated Notification Authority must be provided to the WTO Secretariat. Any changes in this information must also promptly be brought to the attention of the WTO Secretariat. In the appendix a list of countries that have an official National Notification Authority is given.

Enquiry points

Each Member has to ensure that an Enquiry Point exists which is responsible for the provision of answers to all reasonable questions from Members. The Enquiry Point also provides the relevant documents regarding:

- all existing and proposed SPS measures,
- control and inspection procedures, production and quarantine treatment, pesticide tolerance and food-additive approval procedures,
- risk-assessment procedures, factors taken into consideration, as well as the determination of the appropriate level of protection,
- membership and participation in international and regional sanitary and phytosanitary organizations, as well as in bilateral and multilateral agreements and arrangements (including on equivalence), and the texts of such agreements and arrangements.

Full identification of the Enquiry Point must be provided to the WTO Secretariat.

The SPS Committee

The simple existence of transparency obligations under the Agreement does not exhaust the theme. In practice, the governing body of the Agreement, the SPS Committee, functions as a privileged forum for the discussion of health-related trade concerns where WTO Members exchange information on all aspects related to the implementation of the SPS Agreement. The Committee reviews compliance with the agreement and discusses specific trade concerns and all matters related to notification and transparency in general.

Specific trade concerns brought to the attention of the SPS Committee

The Committee normally meets three times a year and all the 144 WTO Members, 22 acceding countries and observers, have the right to attend its meetings. Discussions of trade concerns, especially market-access issues, brought to the attention of the Committee by Members, are central to its work. Since the entry into force of the Agreement on 1 January 1995, more than 115 health-related trade concerns were raised (human or animal health or plant protection). After a slow take-off, the number of concerns raised has risen steadily in recent years and in 2002. Members raised 29 new trade concerns. Importantly, a growing number of trade concerns are raised by developing countries, which is an encouraging sign that the involvement of these countries in the work of the Committee is improving. Additionally, a standing agenda item allows for the discussion of specific notifications. All that WTO Members need to do is to inform in advance the Secretariat and the concerned Member of their intention to raise an issue at the Committee meeting.

This simple and low-cost mechanism not only improves transparency (most often a trade concern is shared by more than one country) but has also proven its efficiency. Several countries have reported breakthroughs in consultations or complete solutions of trade problems as a result of discussions before, during, or after the Committee meetings. Of course, to raise trade concerns at SPS Committee meetings is a useful practice but does not solve all problems. The WTO system provides for a range of alternative approaches to help Members to preserve their rights and obligations. These include consultations and voluntary good offices, conciliation or mediation. However, Members have the right, under the WTO Dispute Settlement Procedures, to request the establishment of a panel at any moment to resolve their trade disputes. Additionally, parties to a dispute have the right to appeal by bringing the panel findings to the WTO Appellate Body.

Appendix.

WTO Members having established an Enquiry Point and/or a National Notification Authority (February 2003)

WTO Member	Has established SPS: Enquiry Point	National Notification Authority*	Has Notified SPS Measures
Albania	X	X	X
Angola[1]			
Antigua and Barbuda	X	X	
Argentina	X	X	X
Armenia			
Australia	X	X	X
Austria	X	EC	X
Bahrain	X	X	X
Bangladesh[1]	X		
Barbados	X	X	X
Belarus[2]	X	X	
Belgium	X	EC	X
Belize	X	X	
Benin[1]	X	X	X
Bolivia	X	X	X
Botswana	X	X	X
Brazil	X	X	X
Brunei Darussalam	X	X	X
Bulgaria	X	X	X
Burkina Faso[1]	X		
Burundi[1]			
Cameroon	X	X	
Canada	X	X	X
Central African (Rep.)[1]			
Chad[1]			
Chile	X	X	X
China	X	X	X
Colombia	X	X	X
Congo, Republic of			
Costa Rica	X	X	X
Côte d'Ivoire	X		
Croatia	X	X	
Cuba	X	X	
Cyprus	X	X	X
Czech Republic	X	X	X
Democratic Rep. of Congo[1]			
Denmark	X	EC	X
Djibouti[1]	X	X	
Dominica	X	X	
Dominican Republic	X	X	X
Ecuador	X	X	
Egypt	X	X	
El Salvador	X	X	X
Estonia	X	X	
European Communities	X	X	X
Fiji	X	X	X
Finland	X	EC	X
France	X	EC	X
Gabon	X		
Gambia[1]	X	X	
Georgia	X	X	X

WTO Member	Has established SPS: Enquiry Point	National Notification Authority*	Has Notified SPS Measures
Germany	X	EC	X
Ghana	X		
Greece	X	EC	
Grenada	X	X	
Guatemala	X	X	X
Guinea Bissau[1]			
Guinea, Rep. of[1]			
Guyana	X	X	
Haiti[1]		X	
Honduras	X	X	X
Hong Kong, China	X	X	X
Hungary	X	X	X
Iceland	X	X	X
India	X	X	X
Indonesia	X	X	X
Ireland	X	EC	
Israel	X	X	X
Italy	X	EC	X
Jamaica	X	X	X
Japan	X	X	X
Jordan	X	X	X
Kenya	X	X	X
Korea	X	X	X
Kuwait	X		
Kyrgyz Republic	X	X	
Latvia	X	X	X
Lesotho[1]			
Liechtenstein	X	X	
Lithuania	X	X	
Luxembourg	X	EC	
Macao, China	X	X	X
Macedonia[2]	X		
Madagascar[1]	X	X	
Malawi[1]	X	X	X
Malaysia	X	X	X
Maldives[1]	X		
Mali[1]	X	X	
Malta	X	X	
Mauritania[1]	X	X	
Mauritius	X	X	X
Mexico	X	X	X
Moldova			
Mongolia	X	X	X
Morocco	X	X	X
Mozambique[1]			
Myanmar[1]	X	X	
Namibia	X	X	
Netherlands	X	EC	X
New Zealand	X	X	X
Nicaragua	X	X	
Niger[1]	X		
Nigeria	X	X	
Norway	X	X	X
Oman	X	X	
Pakistan	X	X	X
Panama	X	X	X
Papua New Guinea	X	X	

WTO Member	Has established SPS: Enquiry Point	National Notification Authority*	Has Notified SPS Measures
Paraguay	X	X	X
Peru	X	X	X
Philippines	X	X	X
Poland	X	X	X
Portugal	X	EC	
Qatar	X	X	
Romania	X	X	X
Rwanda[1]			
Saint Kitts and Nevis			
Saint Lucia	X	X	
Saint Vincent & Grenadines			
Senegal[1]	X	X	X
Sierra Leone[1]			
Singapore	X	X	X
Slovak Republic	X	X	X
Slovenia	X	X	X
Solomon Islands[1]	X	X	
South Africa	X	X	X
Spain	X	EC	
Sri Lanka	X	X	X
Suriname			
Swaziland	X	X	
Sweden	X	EC	
Switzerland	X	X	X
Chinese Taipei	X	X	
Tanzania[1]	X	X	X
Thailand	X	X	X
Togo[1]			
Trinidad and Tobago	X	X	X
Tunisia	X	X	
Turkey	X	X	X
Uganda[1]	X	X	X
United Arab Emirates	X		
United Kingdom	X	EC	X
United States	X	X	X
Uruguay	X	X	X
Venezuela	X	X	
Zambia[1]	X	X	X
Zimbabwe	X	X	
TOTAL MEMBERS = 145	126	118	90
of which LLDCs: 31	9	4	5

* Notifications for member States of the European Communities are made by the EC Commission ("EC")

[1] Least-developed countries: obligations under the SPS Agreement apply as of 1 January 2000.
[2] Observer status

12

Food-safety activities in the World Bank

Cees de Haan[*]

Rationale for the Bank's involvement

The World Bank supports food-safety activities because of their contribution to the World Bank's mandate of poverty reduction. About seventy-five percent of the world's poor live in rural areas, and for those, agriculture is the principal source of income and main engine of growth. However, it is becoming increasingly clear that production of staple foods for the domestic market is not enough to trigger the economic growth needed to get the rural population out of the poverty trap. For sustained poverty reduction, the production of high-value products is highly desirable. However, access to domestic and export markets for such high-value products depends on compliance with food-safety and quality standards. Thus, support for food-safety capacity building is seen as a critical element in achieving the World Bank's mandate.

Background

The World Bank's involvement in food safety is relatively new. It has had, and continues to have, a long involvement in plant and animal health control, but it has only recently become involved in supporting the developing world in capacity building for more specific food-safety improvements in areas such as standard setting, chain management, risk management, traceability, etc. The World Bank is aware, however, that there are many agencies involved in this area, and therefore seeks to get involved in those areas where it feels it has a comparative advantage. These are listed below.

Main intervention areas

Policy dialogue

The World Bank likes to work with public and private partners on specific policy issues defining the efficiency of the food-safety system in developing countries. Some of the issues to be addressed are:

– *Private- and public-sector roles* in food safety, acknowledging that there are externalities and moral hazards involved in food safety which justify public-sector involvement, but acknowledging that there are also strong private-good elements;
– *Economic growth vs. equity priorities*, acknowledging that there are important economies of scale affecting the profitability of food-safety investments, and that

[*] *The World Bank, 1818 H Street, N.W., Washington, DC 20433 U.S.A.*

A.G. J. Velthuis et al. (eds), New Approaches to Food-Safety Economics, 125-127.

there is a danger that small rural producers and processors are crowded out, in particular in countries with strong structural changes in the production and processing sector. Understanding the factors affecting the cost of complying with SPS standards, the distributional effects that they entail and how to alleviate possible negative effects of the concentration are therefore important aspects of the Bank's global and country-specific work.

- *Domestic- or export-market focus.* While the production for export markets is important for economic growth, meeting export requirements is expensive and sometimes the requirements for the export market are not the most appropriate ones for the domestic market, as local food-preparation habits require different standards. Setting domestic standards at the level of OECD requirements can therefore take certain products out of reach of the poor. But, the export market is also a pull factor for domestic demand and standards, and can produce additional income. In any policy dialogue, an analysis first of the trade-offs between uniformly imposing export standards, establishing enclave production for export markets, and focusing on domestic standards, is therefore important.
- *Food-safety vs. food-quality focus.* This links with the public- and private-sector roles issue. Food safety (i.e. hygienic [health] characteristics) is mainly a public good, while food quality (all attributes which influence product value to consumers) is predominantly the private-sector responsibility.

Institution-building support

The World Bank supports the strengthening of public- and private-sector food-safety institutions in the developing world. This can include the setting up of a monitoring capability (e.g. traceability systems, disease surveillance, residue testing), training in risk assessments, disease control and eradication programs and the establishment of disease-free zones. It can also involve training in hazard prevention for small processors and in HACCP models for larger producers. Finally this might include support to private institutions (development of producer/marketing groups) for private certification and quality standards and labeling, combined with public information systems about results of these control systems.

Equipment and infrastructure support

The level and types of investments that the World Bank supports in food-safety capacity building, strongly depends on the level of development and market access of the country concerned. At the lower end of the development continuum, sanitation and water-supply infrastructure are probably the most critical elements. As the level of development increases, laboratory and other disease-control/eradication equipment to address single-source, specific hazards is often required. As the development level increases even more, equipment for HACCP and other mandated standards enabling a farm-to-table approach becomes more important, as well as the support for research to find or adapt new methods of control for important hazards. Infrastructure to support safe handling and processing is sometimes provided under a matching grant system (whereby the private operator puts in a substantial part of the costs, supplemented by a grant through the project). The grant element then acknowledges the public-good element.

World Bank-funded investment operations

A typical World Bank-funded investment operation in food-safety improvement is about US$ 10-15 million, with a total of about US$ 150 million per year invested in food safety. Most food-safety enhancement investments are components of larger rural development operations, focusing on introducing a holistic approach (farm to table), and combining institutional reforms with investments in hardware and software capacity building. In the project preparation, the World Bank relies mostly on their partners (FAO, OIE in particular) for expertise.

Our current research program

The country-focused investment program described above is complemented by methodology and tool-development work, mostly carried out by the central World Bank departments. This focuses currently on economic research on impact and policy options regarding technical barriers to trade, including SPS, and the development of an SPS toolkit for priority setting in policies and investments. In addition the Bank is launching a study on the costs and equity effects of compliance of developing countries with SPS standards, to identify key areas for intervention.

More information can be found in:

Unnevehr, Laurian, and Nancy Hirschhorn, 2000.
Food Safety Issues in the Developing World, World Bank Technical Paper 469, Washington D.C.

13

Food safety and security system in agri-food chains in Japan

Miki Nagamatsu & Yoichi Matsuki

Background

In Japan, the first outbreak of BSE and a series of false indications for meat have resulted in a loss of confidence in food and have increased the distrust of consumers in the administration and agri-food industry. Therefore, it has become an urgent matter to restore the confidence of consumers by establishing a system of scientific inspection, guarantees, risk analysis and traceability of food, and making the information publicly available. In other words, in the entire process "from farm to table", a safe and secured agri-food chain responding to the need of consumers should be established.

Agri-food chains of organic food

In Japan, consumer groups and co-operatives as pioneers in the movement requesting organic agricultural products and safe food have developed safe agri-food chains led by consumers. The development process of organic farming in Japan is briefly presented below.

- The movement of organic farming in Japan started in the period of rapid economic growth when in search of safe food consumers identified problems and cooperated with the producers who carried out the organic farming. This type of chain is the direct linkage between production and consumption.
- Although organic farming began as a grass-root movement, since the domestic market expanded in the late 1980s and a global market for organic food was formulated in the early 1990s, the domestic market (that did not have the standard of organic food) was subject to great confusion.
- The Codex Committee of FAO/WHO adopted the organic standard of crop products in 1999, and in 2001 it added the guidelines for livestock products.
- In adopting the Codex guidelines, the preparation of standards for organic products was required. Thus, the Japanese Agricultural Standard Law (the JAS Law) was amended in July 1999; it introduced the inspection and guarantee system for organic food except livestock products.

Actual situation of organic farming

The actual number of organic farmers in Japan, according to the Census of 2000, was only about 10,000 out of 3.2 million agricultural households in total. Moreover, the organic farmers who are recognized by a third-party organ are a mere 3,000 households. Since the climate of Japan is characterized by high temperature and humidity, insects and diseases tend to break out, thus full-scale organic farming is difficult to conduct. Moreover, due to the tiny size of farmland of individual agricultural households and the high cost of certification that has to be borne by the

A.G. J. Velthuis et al. (eds), New Approaches to Food-Safety Economics, 129-132.

producers, farmers are not keen to acquire the certificates of organic farming. Specific legislation to promote organic farming, like in the EU, does not exist in Japan.

Development of new, diversified, safe and secured agri-food chains in Japan

The "Sanchoku" (the direct transaction of food from production areas) through a partnership with producers ("Teikei"), in search of safe food for consumer groups and cooperatives, has been developed as a unique marketing system from producers to retailers in Japan. At present, almost all marketing enterprises have introduced the "Sanchoku" business, which becomes an important channel in the Japanese marketing system.

A safe and secured agri-food system by producers and entrepreneurs developed from the Sanchoku

Initially the Sanchoku and the Teikei were considered not to require examination/inspection by third parties. This is because of the following three specific features of the Sanchoku/Teikei:

- Production places and production methods are clear;
- Production and raising methods of products are clear; and
- Producers and consumers frequently communicate with each other.

In other words, the Sanchoku/Teikei is considered to be a chain in which producers can trace all information about production, thus traceability is guaranteed.

Such a movement has been developed under the leadership of consumers, but as the number of consumers who seek food safety and security increases, entrepreneurs of the agri-food industry are developing the chains led by consumers. Three examples of such types of chains are presented below.

The Zen-Noh (the National Federation of Agricultural Cooperative Associations) Security System

The first example is the "Zen-Noh" Security System. Zen-Noh is the largest marketing organization of agricultural cooperatives. After a Zen-Noh security system of beef was developed in 2000, it has been expanded to other agricultural products. The system is not necessarily a system of organic farming and organic animal husbandry, but a system in which the traceability is secured by considering the environment and health of consumers. This system has the following three characteristics: (i) to establish the production standards; (ii) to record all production history; and (iii) to make the production history known to the public. Furthermore, as a new project, the system tries to create new environmental indicators so that parents and children of consumers in urban areas learn about the biological diversity in producing areas by a method called "the survey of living things".

The Nichirei "Kodawari" (obsession) food

The next example is a food developed by Nichirei, a large warehouse company as well as a manufacturer of frozen food. Nichirei responds to the need of consumers for high-quality food and develops and markets a private brand called "Kodawari livestock products" (including chickens raised with Chinese medicinal herbs, and natural pork and organic beef). Its five concepts are "safety", "security", "health", "deliciousness", and "consideration for the environment". To meet these concepts, it has built on its own the three frameworks: (i) a "kodawari" standard of livestock products; (ii) establishing traceability; and (iii) a quality-guarantee system by an independent body within the company.

Oisix e-commerce

Oisix is an e-commerce marketing industry of organic and natural food. Its specific features include: (i) to prepare the standards of handling commodities (regarding the applied amount of pesticides and chemical fertilizers, confirmation of production processes, presence of food additives, and taste); (ii) to provide information about commodities as detailed as possible by means of IT (the establishment of traceability, and public availability of information); (iii) its handling commodities are decided by a committee consisting of both food experts (academics) and consumers; (v) to engage in home delivery of milk and in catalogue sales; and (vi) a new entry into agriculture.

The common elements to the above three cases are to make efforts to establish traceability, and to introduce a quality-assessment method by third parties that can conduct an objective assessment, in responding to the increasing needs for safety and security.

New development of food-safety policy

Laws relating to food safety currently include: (i) the Animal Infectious-Diseases Control Law; (ii) the Feed-Safety Law; (iii) the JAS Law, those being under the Ministry of Agriculture, Forestry and Fisheries (MAFF); (iv) the Law Concerning Meat-Disposing Places; (v) the Law Concerning Food Hygiene; (vi) The Nutrition-Improvement Law, those being under the Ministry of Welfare and Labor (MWL); (vii) the Measuring Law under the Ministry of Economy and Industry; and (viii) the Law Concerning Gift Indication under the Fair-Trade Commission.

Among them, the JAS Law is involved in the indication of qualities and standards of agricultural products. The compulsory indication of organic agricultural products according to the Codex guidelines has been met by amendment of the JAS Law. The indication of product origins applied to genetically engineered agricultural products is also made in accordance with the JAS Law.

However, as it becomes evident that these laws, divided along the lines of several concerned ministries, could not cope with the occurrence of BSE and a series of false indications of food, it is an urgent issue to establish an organ that comprehensively controls the process from production to consumption, and to enact the related laws. Therefore, the Government of Japan decided to establish a food-safety agency. The agency should be independent from MAFF and MWL, but under the supervision of the Prime Minister's office. The agency should be responsible for risk assessment and risk communication, but for the risk management the related ministries (MAFF and MWL) are responsible.

One of the distinct features of the new body is that it is comprised of five experts, but no representative of consumers is included. It requires several hundred public officials and the budget will be allocated through the concerned ministries. The new law will be enacted next year, and the existing Law Concerning Food Safety and the JAS Law will then be amended.

Conclusion

Apart from the establishment of a new administrative body and the enactment of new law, which are urgently required, it is necessary to make more efforts to develop agri-food chains led by consumers. For this purpose, the agri-food chain must be capable to carry out the risk analysis, the food indication system must be transparent and clear, and the traceability of food must be ensured. Furthermore, the consumer-led agri-food system should develop an autonomous standard through the partnership among producers, food industry and marketing entrepreneurs. Above all, the autonomous standard should cover all the stages from farm to table.

List of participants

Bogers, Rob	Wageningen University and Research Centre, Wageningen, The Netherlands
Davison, Ann	European Economic and Social Committee, Brussels, Belgium
De Haan, Cees	World Bank, Washington D.C., United States
De Wit, Wim	Netherlands Food Authority, The Hague, The Netherlands
Den Hartog, Leo	Nutreco R&D, Boxmeer, The Netherlands
Gilbert, John	Central Science Laboratory, Great Britain
Henken, André	National Institute of Public Health and the Environment (RIVM), Bilthoven, The Netherlands
Henson, Spencer	University of Guelph, Guelph, Ontario, Canada
Hogeveen, Henk	Wageningen University, Wageningen, The Netherlands
Huirne, Ruud	Wageningen University, Wageningen, The Netherlands
Jensen, Helen	Iowa State University, Ames, Iowa, United States
Lund, Mogens	Danish Research Institute of Food Economics, Frederiksberg C, Denmark
Magalhães, João[1]	World Trade Organisation, Geneva, Switzerland
Marette, Stephan	INRA-INAPG, Grignon, France
Matsuki, Yoichi	Nippon Veterinary and Animal Science University, Tokyo, Japan
Moen, Ad	Animal Health Service, Deventer, The Netherlands
Meuwissen, Miranda	Wageningen University, The Netherlands
Mueller, Walter	Federal Office for Agriculture, Waedenswil, Switzerland
Nayga, Rudy	Texas A&M University, College Station, Texas, United States
Novoselova, Tatiana	Wageningen University, Wageningen, The Netherlands
Oskam, Arie	Wageningen University, Wageningen, The Netherlands
Oude Lansink, Alfons	Wageningen University, Wageningen, The Netherlands
Riepma, Wim	Product Boards for Livestock, Meat and Eggs, Zoetermeer, The Netherlands
Roosen, Jutta	Catholic University of Louvain, Louvain, Belgium
Schiefer, Gerhard	University of Bonn, Bonn, Germany
Shogren, Jason	University of Wyoming, Laramie, Wyoming, United States

[1] Did not attend, but is author of a paper that was presented by Gretchen Stanton

Smarzcz, Raphaela	Gerhard Mercator University, Duisberg, Germany
Spiller, Achim	University of Göttingen, Göttingen, Germany
Stanton, Gretchen	World Trade Organisation, Geneva, Switzerland
Stärk, Katharina	Swiss Federal Veterinary Office, Bern, Switzerland
Stark, Jacqeus	DSM Food Specialities, Delft, The Netherlands
Tarrant, Vivion	The National Food Centre TEAGASC, Dublin, Ireland
Unnevehr, Laurian	University of Illinois, Urbana, Illinois, United States
Vaarkamp, Annemarie	Wageningen University, Wageningen, The Netherlands
Valeeva, Natasha	Wageningen University, Wageningen, The Netherlands
Van Boekel, Tiny	Wageningen University, Wageningen, The Netherlands
Van Rie, Jean-Paul	Wageningen UR RIKILT - Institute of Food Safety, Wageningen, The Netherlands
Velthuis, Annet	Wageningen University, Wageningen, The Netherlands
Verbeke, Wim	Ghent University, Gent, Belgium
Wagenaar, Jaap	Wageningen UR ID-Lelystad – Institute for Animal Science and Health, Lelystad, The Netherlands
Wilpshaar, Henny	Wageningen University, Wageningen, The Netherlands
Wilson, Norbert	OECD Directorate for Food, Agriculture and Fisheries, Paris, France
Zachariasse, Vinus	Wageningen University and Research Centre, Wageningen, The Netherlands